Hair of West-European mammals

Hair of West-European mammals

B.J. TEERINK

Research Institute for Nature Management, The Netherlands

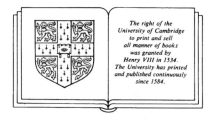

The right of the
University of Cambridge
to print and sell
all manner of books
was granted by
Henry VIII in 1534.
The University has printed
and published continuously
since 1584.

CAMBRIDGE UNIVERSITY PRESS

Cambridge

New York   Port Chester   Melbourne   Sydney

PUBLISHED BY THE PRESS SYNDICATE OF THE UNIVERSITY OF CAMBRIDGE
The Pitt Building, Trumpington Street, Cambridge, United Kingdom

CAMBRIDGE UNIVERSITY PRESS
The Edinburgh Building, Cambridge CB2 2RU, UK
40 West 20th Street, New York NY 10011–4211, USA
477 Williamstown Road, Port Melbourne, VIC 3207, Australia
Ruiz de Alarcón 13, 28014 Madrid, Spain
Dock House, The Waterfront, Cape Town 8001, South Africa

http://www.cambridge.org

First published 1991
First paperback edition 2003

A catalogue record for this book is available from the British Library

Library of Congress cataloguing in publication data

Teerink, B. J.
Hair of West-European mammals/B. J. Teerink.
   p.   cm.
Includes bibliographical references and index.
ISBN 0 521 40264 6 hardback
1. Mammals–Europe–Identification.   2. Hair–Identification.
3. Hair–Atlases.   I. Title.
QL726.T44   1991
599.094–dc20   91-7576   CIP

ISBN 0 521 40264 6 hardback
ISBN 0 521 54577 3 paperback

Contents

Contents

Acknowledgements

I am indebted to Dr C. Smeenk and Mr D. Reeder of the National Museum of Natural History, Leiden for the utilization of the collection of the Museum.

The valuable suggestions of Dr M.H. den Boer after critical reading of the manuscript are deeply appreciated, furthermore my thanks are due to him for correcting the English text.

Finally I wish to express my thanks to Mr R.F. van Beek for developing the films and printing the negatives and to Mrs T. Oostenbrug for typing the manuscript.

1 Introduction

Mammalian hair plays a variably important part in several branches of human society. Since historical time, the hair of mammals has been important not only as fur for clothing but also, in combination with natural or artificial fibres, for the production of cloth. Furthermore, the natural character of the hair of, for example, the marten, badger, and squirrel, is widely used in the manufacture of many different kinds of brushes. Finally – and this brings us to the subject of the present study – scientific research in such fields as animal ecology, wildlife biology, gamekeeping, hunting, and nature management, as well as forensic research, is often supplied with information by the unknown former bearer of hair found in nature or in the stomach, gut, and the faeces of carnivores and the pellets of owls and raptors (Twigg, 1975), where the hair often shows relatively little or no damage. The results of such research often contribute to a more detailed understanding of the distribution of mammals and give the investigator, interested in carnivores, more insight into the quality of the diet of these animals in different areas.

According to Tupinier (1973), early studies on mammalian hair were performed by Brewster in 1837 and Quekett in 1844. In 1920, however, Hausman promoted morphological hair research by undertaking some classification of the great variety of morphological structures occurring in mammalian species, and the terminology he introduced is widely used. Wildman (1954), too, devoted an extensive and thorough study to various kinds of hair, especially in relation to the textile industry. He examined both hair growth and hair morphology, and dealt not only with the manufacture of cross-sections, medulla, and cuticula, but also enlarged the terminology used for the description of cuticular and medullar patterns.

Many other scientists have contributed, each in his own way, to our knowledge of mammalian hair. Lochte (1938) dealt with the hair of a variety of species – up to the leopard, puma, and gorilla – in an extensive atlas. Unfortunately, an appreciable number of European mammals were omitted. Because it was published in the Polish language, the accessibility of Dziurdzik's (1973) key is very limited, but some years later (1978) her study on the Gliridae (six species) appeared in English. In Switzerland, a number of studies dealing with hair were performed. Tupinier (1973) used the scanning electron microscope to examine the cuticle of 29 West-European Chiroptera species belonging to ten genera and concluded that the cuticular scale forms have limited taxonomic value. Keller (1978–1981 *a*, *b*) published keys in a series of four papers providing valuable information concerning hair characters. Although it has no keys and almost no commentary, the atlas produced by Debrot *et al.* (1982) contains drawings of cross-sections and photographs of cuticula and medulla of European mammals that are very useful for comparison. All of these publications have the most relevance for the morphology and structure of the hair of European mammals, but outside Europe, too, a number of scientists contributed to this field: Mathiak, 1938*a*, Williams, 1938, Benedict (1957; Chiroptera), and Brunner & Coman, 1974.

In general, identification of a plant or animal is difficult if only minor parts are available. When there are a few hairs of a mammal, the only clues are provided by the colour, form, and length of the hair and the structure of cuticula, medulla, and cross-sections. Nevertheless, it is often possible to reach correct determination by using combinations of these few data. This makes it absolutely necessary to provide the reader with an identification key in a way allowing him to learn to distinguish the often small differences between hairs. A photo atlas alone is not sufficient for reliable identification. The same holds for the great majority of the published identification keys, because the character information is inadequate and poorly defined. It is especially important that keys intended for hair identification provide both detailed pictures and text to keep the user from going astray. In the present publication he will find detailed keys with many drawings and a comprehensive photo atlas given in Parts II and III, respectively.

2 Hair growth

Several authors have described the development and growth process of hair (Hausman, 1930; Wildman, 1954; Lyne, 1966). The following is a simplified summary of this process.

The skin of an animal has two main layers: a lower layer called the dermis and an upper layer called the epidermis (Fig. 1). The epidermis is composed of layers of dead and living cells. One of the latter is the basal layer whose cells continually divide and maintain the epidermis. The outermost layer of the epidermis forms a thin, horny covering composed of dead cells which peel off continuously. At a given moment, the basal layer starts to show a shallow downward growth locally into the dermis, where, in combination with a dermal papilla, it forms a small solid plug. This is a young hair follicle. Cells of the dermis then form a connective-tissue sheath and small blood capillaries around the growing follicle to supply nutriment to the newly organized tissue

Fig. 1.
Schematic representation of section of a hair follicle.

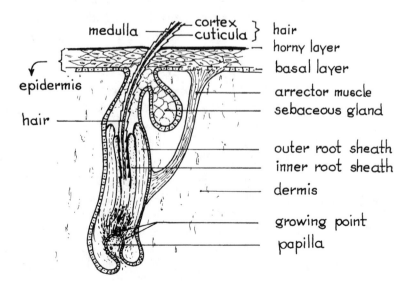

2

2. Hair growth

(Wildman, 1954). Next, outgrowths appear on the neck of the follicle and form a wax gland. In some animal species, an arrector muscle develops as well, which permits hairs to be set up. The cells at, and around, the base of the follicle (i.e. the growing point) then divide rapidly. Those situated beside the growing point form the layers of the inner and outer root sheaths, the hair itself being formed of the more centrally placed cells.

When the cells are pushed further from the growing point, their nuclear bodies gradually become smaller and the formation of horn or keratin starts in the cell. Particularly in the cells forming the inner root sheath, this process of cornification proceeds faster and starts lower down, than it does in the more central layers that form the final hair. Two factors, i.e. the presence of the comparatively rigid inner root sheath round the soft, flexible young fibre cells and the direction of the growing point, play an important role in determining the shape of the hair when it eventually emerges from the skin (Wildman, 1954). The wax gland (called the sebaceous gland) introduces into the young follicle wax cells through which the tip of the growing hair can easily push its way. This, in turn, produces a bulge in the rather tough horny layer of the epidermis still lying over the tip. When the pressure becomes too strong, the bulge in the horny layer ruptures, which allows the tip of the new hair to emerge above the skin surface (Wildman, 1954) (Fig. 1).

A hair consists of three layers, all of which are very important for identification, i.e. the cuticula, the cortex, and the medulla (Fig. 2). In many species the hair follicles are grouped together, the size and arrangement of the groups varying widely. Some time after the onset of follicle formation, two more follicles appear, one on each side of the primordial follicle, forming what is called a trio group. After that, other hair follicles may develop. In various groups of animals, secondary follicles emerge via bulges on the inner side of a hair follicle after passing through roughly the same developmental stages as the original follicle and then using the same opening to emerge. In adult Merinos, for example, as many as nine follicles can occur in one bundle with a common orifice (Lyne, 1966).

The continuation of the physiological processes in the animal's skin finally leads to moulting. Many hair physiologists have discussed this phenomenon. Apparently, the moulting process differs between species. Becker (1952) reported that, in the brown rat, *Rattus norvegicus*, moulting is independent of season, both in the laboratory and in the wild. In *Microtus* and *Apodemus*, moulting takes place throughout the year with peaks in the autumn and spring, especially in *Apodemus* (Stein, 1960). In the bank vole, *Clethrionomys glareolus*, less than 2% of the animals moult in the winter as against 37% in the summer. Moulting is somewhat more dependent on the season in this species (Stein, 1960). In the striped field mouse, *Apodemus agrarius*, the hair is shorter and thicker in the summer than in the winter (Haitlinger, 1968 *a*). In Soricinae, the H-profile is always the same whatever the season (Keller, 1978; Vogel & Köpchen, 1978).

Fig. 2.
Schematic representation of a cross-section of a hair.

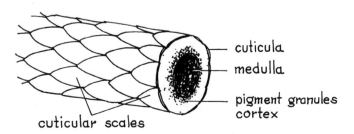

cuticula
medulla
pigment granules
cortex
cuticular scales

Thus, it is clear that moulting does not simplify identification. However, the morphology of cuticula, medulla, and cross-section is not so variable as to make identification impossible.

> *Note*
> During the development of a hair, the proportion outside the skin increases steadily. The structure of the cuticula, cortex, and medulla of the exposed part no longer changes, because the cells have died and become keratinized.
> The fur of a moulting animal shows many hairs in different stages of development, which means that such hairs lack basal parts to some extent. It is of great importance that these parts be distinguished, which can be done on the basis of the cuticular pattern and the medulla. In a full-grown hair the last cuticular cells to be formed always have a simple structure at its extreme base and the medulla is absent in that area (Fig. 3).

Fig. 3.
Basal part of a hair.

extreme base ↑

There are various types of hair, e.g. vibrissae (whiskers), bristle hairs (domestic pig), overhairs, and underhairs, more than one type often occurring on the same animal. Vibrissae are readily recognized, their structure being very similar in different species. Most vibrissae are circular in cross-section, and they are the only hairs to taper all the way from the base to the tip. Since the value of this type of hair for identification purposes is limited, it is not included in the keys. The spines of the Western hedgehog are greatly enlarged and strongly modified hairs. Possible interspecies differences due to age, diet, disease, environment, or season, will not be taken into consideration here.

Due to the absence of GH 1 and GH 2 (see 4.1) in very young animals, identification of these is almost impossible.

3 Geographic area and species studied

This study covers all wild terrestrial mammals of the western part of Europe – including Denmark, Germany, The Netherlands, Belgium, Luxembourg, the northern part of France, Great Britain, and Ireland – except for a few species running wild in Great Britain, sika deer (*Sika nippon* (Temminck)), muntjac (*Muntiacus muntjac* (Zimmerman)), reindeer (*Rangifer tarandus* (L.)), and the genet (*Genetta genetta* (L.)), which is seen occasionally. The agricultural domestic mammals such as the horse, cow, pig, and sheep are not dealt with. A list of the animals included, with their scientific names and common names in English, German, French, Danish, and Dutch, is given at the end of this book.

Part I
How to identify hairs

4 Features with importance for identification and terminology

4.1 Hair profile

As already mentioned, an animal's coat is composed of several types of hair, the main components being the overhair and the underhair (Fig. 4).

Overhair is the long and stiff hair (called guard hair) with a thickening in the distal part called the shield and a thinner proximal part, the shaft (distal = toward the tip, proximal = toward the base of a hair). Underhair is made up much thinner and less stiff and has an undulating appearance. Although the transition from the longest overhair to the shortest underhair is more or less gradual, four main groups can be distinguished, three of them belonging to the overhair (GH 0, GH 1, and GH 2). The scarce GH 0 is a long, firm, and, as a rule, straight hair with an elongated, sharp tip. The diameter of the thickest part of the shield is generally smaller than that in GH 1 and GH 2. This type of hair is most characteristic in the Rodentia (Fig. 4).

The GH 1 hair is usually straight and stiff too but its occurrence is much more frequent (by a factor of some dozens) than that of GH 0. Moreover, the distal part of the shield is less elongated than in GH 0, and this means that the thickest part of the GH 1 shield is a little closer to the tip (Fig. 4). In some species, GH 1 may be slightly wavy or even bent (see species numbers 10 and 49 in the atlas section).

In GH 2, the shield and shaft usually form an angle with each other. The shaft is rather straight as a rule, but may be wavy or zigzag in some groups and/or species (Figs. 56a and 60; species number 55 in atlas section). In other species, GH 2 is rather similar to GH 1, and, in such cases, the smaller type of GH 1 is classified as GH 2. This often occurs in the somewhat larger animals such as the badger (*Meles meles*), wild boar (*Sus scrofa*), and deer (*Cervidae*). The GH 2 type is always more numerous than GH 1.

Fig. 4.
Types of hair (the tip of the hair points to the left).

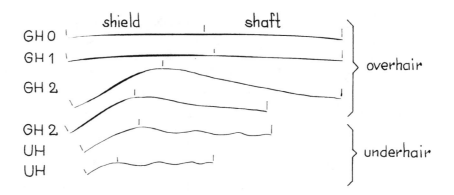

Underhair is the predominant type of hair in a coat but has limited taxonomic value. The shield, the distal articulation of this type of hair is insignificant and shows little or no thickening (Fig. 4).

4.2 Cuticula

The outermost layer, the cuticula (Fig. 2), is made up of a large number of overlapping transparent scales of keratin. The distal part of each scale lies over the proximal part of the next one, and, as a result, each hair has less resistance from base to tip than in the opposite direction. This property is useful for rapid determination of which end of a hair is the tip. When a hair is taken between the thumb and forefinger of each hand and pulled gently, the tip of the hair will be on the end that slips out of the fingers.

The size and shape of the scales vary according to the position on the hair. At the tip, the first scales to appear are small. On the widest part of the shield, the scales are much larger and lie transversely (Fig. 5). Many different shapes occur along the shaft. In general, these scales represent differences between species and/or groups better than those in the shield do. The last scales to be formed, i.e. those at the extreme base of the hair, are simple in form (Fig. 3), and are very similar in all species. The presence of this type shows that the growing stage has been completed (see under Note in Chapter 2).

The shape of the cuticular scales is very important for identification, but usually they can not or can hardly be seen with a light microscope without special preparation. This problem can be solved by making a gelatin impression of the hair (see Chapter 5 under *Material and techniques*), in which the outline of the scales can be seen as thin lines together forming what is called the scale pattern.

4.2.1 Scale position relative to the longitudinal axis of the hair

Transversa

These scales lie at right angles to the longitudinal axis and their width is greater than their length (Fig. 5, also Figs 8, 13, 15, 17, 18).

Longitudinal

The scales lie parallel to the longitudinal axis of the hair and their length is greater than their width (Fig. 6, also Figs 9 and 10).

Intermediate

In these scales the length is approximately the same as the width (Fig. 7, also Figs 11 and 12).

4.2.2 Cuticular patterns

Petal pattern

The general appearance of this pattern is similar to that formed by the overlapping petals of a flower (Wildman, 1954).

   1. Broad petal pattern. This pattern is made up of wide scales (Fig. 8).
   2. Elongate petal pattern. This pattern is intermediate between the broad and the diamond petal patterns (Fig. 9).

Diamond petal pattern

In this pattern the scales overlap in a way giving them a diamond shape. The pattern resembles that of a pine cone.

   1. Narrow diamond petal. The scales are long and narrow (Fig. 10).
   2. Broad diamond petal. The scales are rather short (Fig. 11).

4. Important features and terminology

Mosaic pattern
The adjacent scales have rather straight margins (Fig. 12).

Waved pattern
The pattern is usually transverse with weakly to strongly undulating margins.
We distinguish:
1. The regular wave: The scale always lies transversely. The waves are shallow (Fig. 13).
2. The irregular wave: The scale usually lies transversely (Fig. 5) and sometimes longitudinally (Fig. 6). The waves have deeper troughs and are less regular (Fig. 14).

Figs 5–22.
Cuticular patterns
(the tip of the hair points to the left).

I SCALE POSITION IN RELATION TO LONGITUDINAL DIRECTION OF THE HAIR

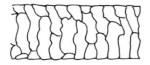
5. transversal

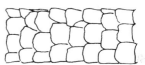
6. longitudinal

7. intermediate

II SCALE PATTERNS

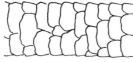
8. broad petal

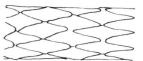
9. elongate petal

10. narrow diamond petal

11. broad diamond petal

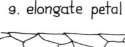

12 mosaic

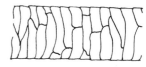
13. regular wave

14. irregular wave

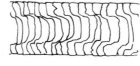
15. streaked

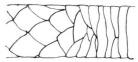
16. transitional

III STRUCTURE OF SCALE MARGINS

17. smooth

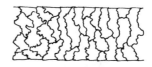
18. rippled

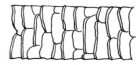
19. frilled

IV DISTANCE BETWEEN SCALE MARGINS

20. distant

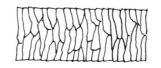
21. near

22. close

3. The streaked pattern: This occurs in hairs with a longitudinal furrow in the shield (Fig.15).

Transitional pattern
Occurs between two patterns (see, e.g. Fig. 16), frequently occurs just distal to the lowest part of the shaft. A similar alteration can always be seen in the proximal part of the shield.

4.2.3 Structure of scale margins

Smooth: The margins of the scales show no indentations and appear as a smooth line (Fig. 17).

Rippled: There are small indentations along the margins, usually close together (Fig.18).

Frilled: This term is used for scales showing narrow borders ($\pm 1\ \mu$m) along the distal part. Such borders may be rather smooth or dentated (Figs 19, 198a, 200b). The width of the border may increase up to 3 μm toward the tip in some species (Figs 198b and c), and in such cases half or more of the next scale may be covered.

4.2.4 Distance between scale margins

The distance between scale margins may vary considerably. To indicate such differences, the terms distant, near, and close (Figs 20–22) have been introduced.

4.3 Cortex

The central layer between the cuticula and the medulla, i.e. the cortex (Fig. 2), has been discussed by Hausman (1930, 1932). It is composed of longitudinal, cornified, and shrunken cells, which appear under the light microscope as a homogeneous, hyaline mass without any details, and is therefore of limited value for identification. In some cases, however, the thickness of the cortex relative to the total width of the hair can be important. Here, use will be made of the term CC/TW (expressed in μm), in which CC is the width of the combined cuticula and cortex and TW is the total width of the hair. Both CC and TW must be measured in the widest part of the shield (Fig. 96), using the medullar slide.

The pigments which are largely responsible for the hair colour may take the form of discrete granules, large amorphous masses, or diffuse stain (Hausman 1930). The majority of the granules are found in the cortex, whereas the other forms occur almost entirely in the medullary column. Because of the large variation, even within the same type of hair, the pigment granules are of little value for identification.

4.4 Medulla

The pith of a hair, the medulla (Fig. 2), is composed, like the cortex, of closely packed, shrunken dead cells, but unlike the cortex they are clearly visible. These cells and the air-filled spaces between the intercellular connections are responsible for the specific character of the medulla.

In many cases, the medulla is so dark that structures can hardly be seen. Substantial improvement can be obtained by saturation of the hair with paraffin oil; this fills the air spaces, which become transparent (for technique, see Chapter 5). For identification, slides both with, and without, oil penetration are needed. In general, the medulla on the widest part of the shield is the most important structure, especially for identifying the orders. Morphologically, the medulla shows many different patterns. Some of them are very distinct, whereas others, for instance those formed by structures in the vicinity of the outer layer of the medulla in some carnivores, can only be seen after careful inspection.

The main patterns and the related terminology are discussed in the following section. In medullar drawings and photographs with both light and dark parts, the light area represents the situation after oil penetration and the dark part the situation without oil penetration.

> *Note*
> Sometimes the digestion process in the stomach and gut of a predator changes the medulla of the prey to a degree making identification of the original pattern difficult, but experience usually solves most of these problems.

4.4.1 Composition of the medulla
The cellular structure of the medulla can be unicellular or multicellular.
  A. *Unicellular*: The medulla is composed of a single layer which is one cell wide, the pattern being regular (Fig. 23) or irregular (Fig. 24).
  B. *Multicellular*: The medulla is composed of two or more layers of cells (Figs 25 and 26).

4.4.2 Structures of the medulla
The arrangement of the cells forming the medulla can take the following forms:
  A. *The ladder pattern*: Between all cells there is a lighter area with almost the same dimensions, which creates a ladder effect (Figs 23 and 27). This pattern is usually restricted to the shaft. Sometimes there is a pattern resembling letters X, Y, V, M, N, or reversed (Fig. 24), which is brought about by air-filled spaces lying obliquely between contiguous cells.
  B. *The intermediate pattern*: A ladder pattern is sometimes so indistinct that a sort of 'wreathy' pattern called intermediate arises. This phenomenon usually occurs in the shaft (Fig. 28).
  C. *The cloisonné pattern*: Shrinking of the cells forms thin, intersecting thread-like lines giving rise to an angular network pattern called cloisonné. Within these partitions, some structured elements can be clearly observed. The spaces within these structures are transparent and are slightly lighter because of the absence of structure (Fig. 29, right side shows oil penetration).
  D. *The reversed cloisonné pattern*: In this type of medullar pattern the cells are voluminous and the spaces between them form a thread-like pattern. The cells are granular, the thin spaces between them lack visible structure (Fig. 30). This pattern is multicellular but the cells do not lie in rows and the pattern often lacks strong contrasts.
  E. *The isolated pattern*: The dark cells are occasionally contiguous but are usually separated to a variable degree and are easily recognized. The shape is circular up to oblong (Figs 25 and 31).
  F. *The crescent pattern*: The dark cells form a pattern imposed by their shape. The cells are rather long and slightly curved, tapering at the ends. Many of them touch and overlap each other. The spaces between the curved cells have the shape of a banana or crescent (Fig. 32).
  G. *The filled pattern*: Medullar cells seem to fill the entire width of the hair,

and the cortex can hardly be distinguished (Fig. 33). The pattern is
multicellular, but the cells do not lie in rows. A cloisonné becomes very
distinct after the penetration of oil (Fig. 277).

H. *The interrupted pattern*: This pattern takes its name from the absence
of the medulla at one or more sites (Fig. 34). If several, separate little
pieces of the medulla are present, the pattern is called fragmental.

4.4.3 Margins of the medulla

Certain features of the margins of the medulla have importance for
identification. These features can be seen most clearly before treatment of the
hair with oil and concern only the widest part of the shield.

Figs 23–37.
Appearances of the medulla
(the tip of the hair points to
the left).

I WIDTH COMPOSITION OF THE MEDULLA

23. unicellular, regular

24. unicellular, irregular

25. multicellular

26. multicellular in rows

II STRUCTURE OF THE MEDULLA

27. ladder

28. intermediate

29. cloisonné

30. reversed cloisonné

31. isolated

32. crescent

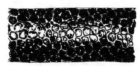

33. filled

34. interrupted

III FORM OF THE MEDULLA MARGINS

35. straight

36. fringed

37 scalloped

> A. *Straight margins*: The margins of the medulla form a smooth, straight line (Fig. 35).
> B. *Fringed margins*: Small protrusions extend into the cortex (Fig. 36).
> C. *Scalloped margins*: A series of convex, rounded projections form the margin of the medulla (Fig. 37).

4.5 Cross-sections

The shapes of hair show considerable variation, ranging from the simplest cylinder (Fig. 38) to highly complex forms (Fig. 50). The numerous, longitudinal ridges and grooves present in hairs of some species may serve to hold air thus providing insulation against extreme air temperatures and used during diving and swimming. Hutterer & Hürter (1981) established an increase of the number of longitudinal ridges related to an increasingly aquatic life of shrews.

The shape and dimensions of hair in cross-sections are very important features for hair identification. Generally, these features are most characteristic in the widest part of the shield. However, cross-sections taken at other positions along the hair are also very valuable if the positions on the hair are known.

Many pictures accompany both the atlas and the keys for purposes of comparison. The most important shapes are depicted in Figs 38–50, together

Figs 38–50.
Forms of cross-sections.

CROSS-SECTIONS

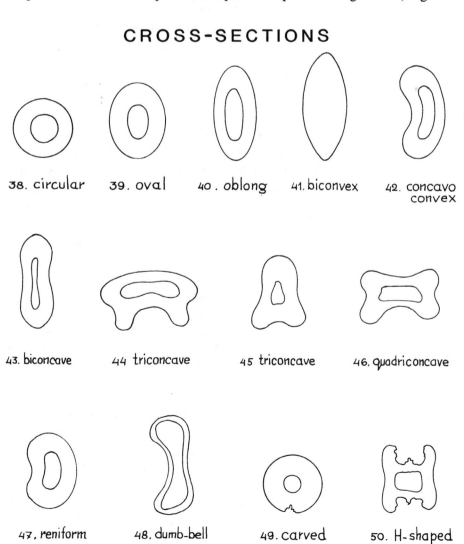

| 38. circular | 39. oval | 40. oblong | 41. biconvex | 42. concavo convex |

| 43. biconcave | 44 triconcave | 45 triconcave | 46. quadriconcave |

| 47. reniform | 48. dumb-bell | 49. carved | 50. H-shaped |

with their terminology. The cross-sections and their medullas are shown without indication of pigment.

5 Material and techniques

5.1 Hair type and stage of development

In the atlas and the keys only the full grown GH 1 and GH 2 (see 4.1) occurring on the animal's back are taken into consideration (Fig. 4), since they show the most characteristic features. Lateral and ventral GH 1 and GH 2 generally carry similar characteristics but often less distinctly: they are sometimes smaller and the colour is usually brighter. GH 0 (Fig. 4) is omitted because of its scarcity and the fact that the cross-sections are usually circular. The same holds for specific types of hair, e.g. hair of the tail and the head, foodtufts, whiskers, and eyelashes.

Although UH is the most common type of hair in a pelage (Fig. 4), it is of little value for identification. In general, four features of GH 1 and GH 2 are important for identification:

1. The cuticula in shaft and proximal shield.
2. Cross-sections through the shield.
3. Medulla in the thickest part of the shield.
4. Medullar margins in the thickest part of the shield.

Because a given type of hair of a species may show some variation, several slides of each type of hair should be made.

5.2 Preparation of microscopical slides

5.2.1 Precautions

For good-quality slides, the hairs must be clean and free of grease. For this, they have to be washed in tepid water containing a detergent, rinsed in water and distilled water, and, if necessary, stored in 70% alcohol. Pieces of a fur can be similarly treated. After drying, immersion in an ether or carbon tetrachloride bath will remove the fatty wax layer.

The most convenient way to perform the procedures of washing and rinsing hair is to place it in a fine-meshed sieve, in which it can be transported from one bath to another without any danger of loss.

> *Note*
> 1. Pelts can be conserved dry by scraping the grease off the surface of the inner side. After that, the pelt is thoroughly washed, rinsed, rubbed with alum, and then stored with naphthalene when it is dry.
> 2. If only a few hairs are available, a single hair can be used for several observations. Slides should be made of cuticula first, then medulla, and finally cross-sections.

5.2.2 The position of the hair in slides and pictures

In the keys and the atlas, hairs are consistently described and depicted with the tip to the left and lying flat.

In slides of cuticula and medulla the flat side of the shield is placed against the glass (Fig. 51a). The more or less symmetrical structure of the shield means that the hair can be placed on the glass in the most convenient way, i.e. with the smallest angle between the shaft and the shield facing away from the glass (Fig. 51b).

For profile slides, the hairs have to be allowed to retain their natural shape. For cross-sections, positioning on the slide is irrelevant.

5.2.3 Hair profile slides

The hair is placed on an object glass such that the profile stands out as clearly as possible. Next, without the use of any mounting medium, a coverglass, provided with a small drop of glue on each corner, is placed over the hair. The best observations are made with a stereomicroscope.

5.2.4 Cuticular slides

As a rule, the surface structure or scale pattern of the cuticula cannot be properly observed without the use of special techniques. To study the structure of the cuticula, some investigators use the scanning electron microscope (SEM) (Vogel & Köpchen, 1978; Dziurdzik, 1978; Short, 1978; Hutterer & Hürter, 1981), which permits direct observation at high magnification giving beautiful images, but this microscope is not obligatory for identification. For this purpose, several other methods are available (Wildman, 1954; Williamson, 1951). With all of them a cast is made in a suitable medium (gelatin, polyvinyl acetate, celluloid, nail polish). The shape of the cuticular scales is transferred to the medium, where it can be observed with a microscope.

The procedure described below, using gelatin as medium, is easy to perform and gives excellent results.

A 10–20% gelatin stock solution is prepared with distilled water to which a few crystals of phenol or thymol have been added as preservative. The mixture is softened by heating in a water bath. A thin layer is brought onto a coverglass of suitable size. The best way to accomplish this is to apply some warm gelatin over the entire coverglass and pour off as much of it as possible. If the dripping side of the glass is then brushed perpendicularly over a sheet of filter paper, a thin film is left on the glass. When the gelatin has been allowed to cool for about five minutes, it will have set. Next, a pair of tweezers is used to place the selected hairs, one by one, and side by side, with the flat side of the shield facing down, after which some pressure must be applied with the tweezers. After drying for about 30 minutes, the gelatin will be solid and the hairs can be removed. A needle can be used to lift the tip and then the hair can be grasped with the tweezers (avoid damage to the distal 3 mm). If required, the hairs can be kept by fixing them in place next to the impression by tiny droplets of water.

To protect the impressions, the coverglass is glued with a small drop of glue at each of the corners, with the gelatin side lying on an object glass. The combination of twice a negative, i.e. the cast and the position on the glass, yields a positive presentation of the cuticular scale.

The slide can be examined under high magnification.

Fig. 51.
In cuticular and medullar preparations, the flat side of the shield is laid down against the glass.

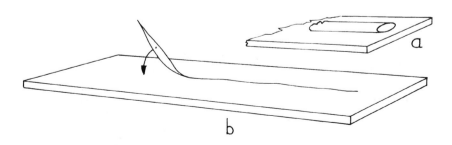

Note
A rapid and easy method to obtain cuticular patterns is to use previously prepared coverglasses on which a gelatin film has already hardened. The hair is placed on the film and the gelatin around it can be softened with a small paintbrush dipped lightly in water. The brush should not be very wet. The gelatin will harden again very quickly, after which the hair can be removed. After the observations have been completed, the impressions can be removed by a stroke of the brush dipped in hot water, and the gelatin film can be used again.

5.2.5 Medullar slides

For identification purposes, the medulla is examined before and after paraffin oil penetration. Paraffin oil completely or partially replaces the air in the intercellular chambers. Dark and opaque medullas usually become transparent under oil penetration, which improves observation of the structure and position of the shrunken cells.

A medullar slide is prepared as follows. The flat shield side of a hair is placed on an object glass, and fixed in place with small drops of white bookbinder's glue (polyvinyl chloride acetate) at a number of points. After the glue has hardened (some minutes), the hair is cut between the glue drops with a sharp razorblade (Fig. 52). Be sure to include a cut at the widest part of the shield. The glue will prevent the hair segments from floating away during mounting. Four or five such hairs can be mounted on one slide. Next, mounting is performed with paraffin oil and a coverglass. To avoid soiling the preparation, the oil should not be allowed to run outside the coverglass. The paraffin oil penetrates the pieces of hair to a variable degree. Because of these differences in penetration, both appearances of the medulla are usually available. If penetration is insufficient, the slide can be put in a warm place or heated gently for some time.

For permanent preservation, the paraffin oil is replaced by, e.g. Canada balsam or dammar resin.

Note
The Chiroptera treated in this study have no medulla at all (see Part III).

5.2.6 Cross-section slides

Several techniques for the preparation of cross-sections of hair are described in the literature. Mathiak (1938*b*) applied a celluloid solution to balsa wood, Wildman (1954) used a special microtome, and Ford & Simmens (1959) used a metal plate with a hole.

With respect to cross-sections, it is essential to know the exact location on the hair because the shape of cross-sections varies with the position on the hair. An entire hair can be viewed throughout the procedure of sectioning if use is made of a transparent embedding medium.

For the method described here, the hairs are mounted between two layers of white transparent cellulose acetate. The procedure is as follows. Take a

Fig. 52.
For medullar preparations the hair is cut between the drops of glue.

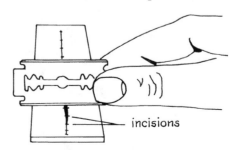

incisions

clean objectglass, use a pipette to apply some acetone along its length, and immediately lay a strip of cellulose acetate (50 × 10 × 0.2 mm or longer if desired) in the liquid starting at one end of the slide. Blot superfluous acetone away with filter paper and press the strip down firmly for a short time. The acetone will dissolve the surface of the cellulose acetate strip, which will attach to the glass. Next, lay a hair lengthwise on the strip and fasten it in place with a small drop of acetone applied at the tip. Lay a thin strip of paper across the hair near the base (Fig. 53). The hair can be stretched by moving the paper with a pair of tweezers. The utmost base of the hair is then fastened in place in the same way as the tip, after which the entire hair is fastened with acetone. Proximally from the base of the hair, bring a sufficient amount of acetone onto the cellulose acetate strip and place a second strip of cellulose acetate on the first one in the same way and without introducing air bubbles. The hair, still clearly visible, will then be mounted between the two strips of cellulose acetate now forming a single strip 0.4 mm thick (Fig. 54). The places to be cut for cross-sections can be marked by making shallow incisions with a razorblade. Before being cut, the strip must be removed from the coverglass with a razorblade. Next, cut the strip lengthwise to a convenient width of approximately 2–3 mm. For proper orientation of the slices later on, cut one side perpendicularly and the other obliquely (Fig. 55c). Put this narrow strip

Fig. 53.
The hair is straightened.

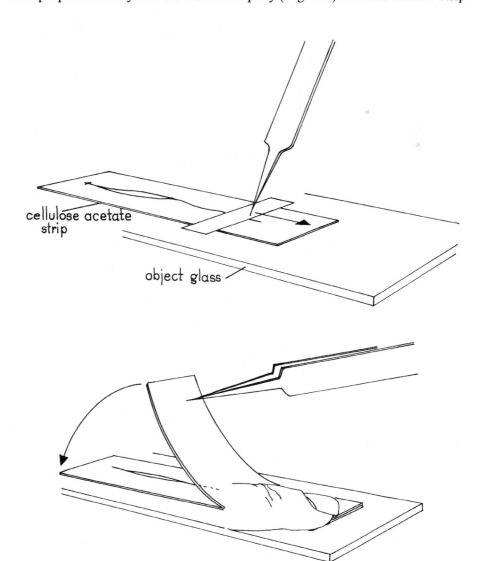

cellulose acetate strip

object glass

Fig. 54.
The second strip is placed.

on the surface of some firm material (plastic, cellulose acetate) and fasten one end in place with adhesive tape (to avoid confusion, always use the same orientation of base and tip). Use a fat-free razorblade to cut cross-sections 50–100 μm thick, by preference under a stereomicroscope, and place the sections on an objectglass. All sections should be similarly oriented; the axis as well as the oblique angle (Fig. 55a and b.)

Before being mounted, the sections must be securely attached to the objectglass. Lay a coverglass such that half of each section is covered (Fig. 55a (left) and d), the other half extending beyond the coverglass. Glue these free ends down with a brush dipped in white vinyl glue diluted with water. The glue must, on no account, run underneath the slices. Then repeat this procedure for the other half of the sections. For mounting, use dammar resin or Canada balsam.

It is advisable to keep the other parts of the hairs and strips together with the cross-sections by attaching them to the same slide in the same sequence. In this way the hair and the locations from which sections were made are always to trace (Fig. 55a (centre) and c). Lastly, label the slide as shown in Fig. 55a (right).

Cross-sections of a bundle of hairs can be prepared in a different way. Lay the bundle on a slide carrying some 70% alcohol. Just before the alcohol evaporates completely, cover the bundle with a small amount of a 3% solution of cellulose acetate in acetone with a pipette. When the material has dried, repeat the procedure once or twice. If the bundle is thick enough, it

Fig. 55.
A coverglass is used to secure the separate sections on the object glass (a,d). The strip is cut obliquely at one length side to know the right position after section (a–d).

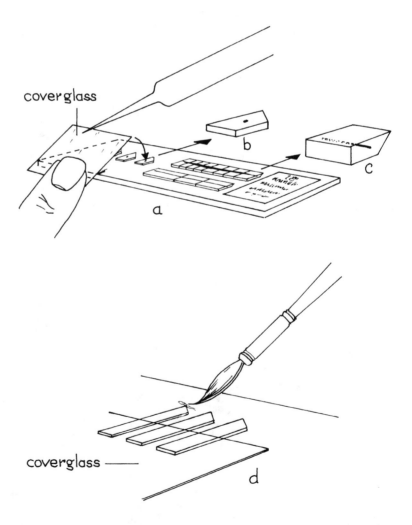

can be sliced as a whole; otherwise, handle it according to the procedure described above.

Note

1. Hairs of some Artiodactyla species can be so soft that they become completely deformed during cutting, but such hairs are so thick that they can be sliced without being embedded. After a small drop of water is applied, they usually regain their original shape.

2. A more rapid but less permanent method to prepare cross-sections for a preliminary impression of shape and size makes use of adhesive tape and cellulose acetate strips of the same size as mentioned above. Paraffin oil is used as mounting medium. The hair is placed on the sticky side of the tape and the whole is glued to the cellulose acetate strip. Most of the air inclusions bordering the hair can be removed by piercing the tape with a needle at several places and rubbing it to help the air to escape. Sectioning is started from the tape side. The slices do not have to be glued to the objectglass as before, because they are now sufficiently sticky to remain in place during mounting.

5.3 A visual and metric device for comparison of dimensions in microscopical images and illustrations in this report

The ratio between a microscopical image and the field size of the microscope changes with the factor of magnification. To visualize this ratio, one of the present Figures can be placed in a circular field of the same ratio. For this purpose a circular hole is cut in a sheet of paper which is then placed over the picture. Depending on the magnification used, the size of the hole can be easily obtained as follows.

The diameter of the visual field of the microscope used must be calculated in ocular micrometer units (OMU). Since the micrometer is located within the ocular tube, it will be independent of the objective and consequently will always have the same dimension. Suppose this to be X. The magnification factor of the microscope is used to calculate the equivalent of one OMU in mm. Suppose the latter to be Y. With a magnification factor in the figure of Z, the diameter of the hole in the paper must by X.Y.Z. millimetre. The higher the magnification factor of the microscope, the smaller the representative value of 1 OMU and hence the diameter of the hole. In general, microscopical slides are observed best with the × 10 ocular and the × 40 objective.

Another useful device for the comparison of dimensions in a preparation with those in the present pictures is an easily made paper measuring rod. This has to be derived from the ocular micrometer in the microscope. The dimensions of the preparation, in OMU are then directly comparable with those of the picture obtained with the derived rod. The measuring rod is obtained in the following way. Suppose, 1 OMU equals X millimetre (magnificaton factor of the microscope). Next, suppose the magnification factor of the picture in question equals Y. One unit on the derived measuring rod must be X.Y. millimetre.

Part II
Keys to groups and species

6 Introduction and standards of magnification

A large number of characters must be used to distinguish between hairs of the different species. Key 1 represents a first classification, which in some cases leads to the species level. In most instances, however, it leads to a group or family which can be further identified in keys 2–12.

Before the keys are consulted, it is recommended that all the required preparations of GH 1 and GH 2 be made and labelled. Due to variability of hair characteristics within a single species, several slides are needed for each type of hair and item. It may be mentioned here once more that in these keys reference is always made to the appearance of the flat side of the shield of the hair in cuticular and medullar slides in both text and illustrations. Dimensions given in the keys should be seen as indication and as support for identification. However, the user must keep in mind that measured values sometimes differ from the stated values. The same holds for drawings and photographs in this article.

For a direct comparison of sizes between microscope images and figures the reader is referred to Section 5.3, where some techniques are described. Drawings of cross-sections are only outlines, without representation of pigment. The number given just before a species name is the permanent number given to that species and is used throughout this paper.

Within orders, an attempt was made to maintain the same magnification for pictures of hair profiles, patterns, and cross-sections in the atlas section to simplify comparison of species. This could not always be done because of the differences in the length and width of hairs and a compromise between clearness (= size) of the pictures and the use of other magnifications had to be made. For both hair profiles and cross-sections, at most two magnifications have been used within the same order. For all cuticular and medullar patterns the standard magnification is 400 except for Artiodactyla, where use is made of the standard in combination with 100. Figures 56–284 accompanying the keys in Part II are all magnified 400 times unless otherwise mentioned.

Note
1. In all the figures the tip of the hair points to the left.
2. Measurements and information about the characteristics of hairs always pertain to the thickest part of the shield unless otherwise mentioned. Both measurements and the data concerning the term CC/TW should be seen as an indication and support for identification rather than as exact measurements.
3. With respect to general patterns in the keys (e.g. as in Chapter 4) the reader

should take into account that these refer only to the appearance of the hair characteristics in question; they do not apply to measured values. This also holds for references to patterns of species other than those concerned.

Key 1   Key to groups of animals and to some species with exceptional hairs

1a.   Hairs greatly enlarged, becoming spines.
Implantation into skin hemispherical (Fig. 180). A × 40 magnification resembling a maize cob; cross-section Fig. 179.

1 *Erinaceus europaeus*
Key 2

  b.   Hairs of the normal type.   2

2a.   Medulla absent, hair length less than 15 mm.   Chiroptera
Key 3

  b.   Medulla present   3

3a.   GH 2 with distinctly zigzag shaft and distinctly enlarged central shield (Fig. 56a).
GH 2 with markedly protruding cuticular scales along one side of the shaft, crossing over to the other side after each curve (to be seen in hair-profile slides, Fig. 56a).
GH 1, if present, straight (Fig. 56b); cross-sections in shield often circular, or with variably sized incisions (Fig. 56c).
In GH 2, this feature is much more pronounced (Fig. 56d–g).
Both GH 1 and GH 2 less than 10 mm long.   Insectivora
Key 2

  b.   Hairs, short or long, straight, curved, curled, more or less zigzag, or even otherwise. In zigzag forms, cuticular scales on both sides of shaft protruding in a similar way and not unequally as shown in Fig. 56.   4

4a.   GH 1 long, up to more than 100 mm, straight, and very coarse, split at tip, often into more than one part (Fig. 279).

69 *Sus scrofa*, see also
Key 12

  b.   Hairs not split at tip.   5

5a.   GH 1 with *c*. 10–15 mm long white tip; proximal to this, a 15–20 mm

dark region, the remaining basal part of the hair being light to white.
Largest cross-sections of GH 1 up to more than 200 μm (Fig. 57).
Combined cuticula–cortex in thickest part of the hair *c.* 45–60 μm.
Cuticular pattern of GH 1 variable in shaft as well as in the larger
type of GH 2 (Fig. 266 or as in Figs 68 and 69). The smaller type
of GH 2 shows a proximal shaft with a specific pattern which can
be reduced to a diamond petal pattern (Fig. 59). Medullar margins
of GH 1 and GH 2 in thickest part of shield almost straight.

65 *Meles meles*
Key 10

b. Otherwise 6

6a. Medulla of GH 1 and GH 2 multicellular with five to eight
longitudinal regular rows of cells (Fig. 26), cuticula in central and
distal parts of shaft often with pattern as shown in Figs 206 and
208, respectively.
Length of GH 1 25–40 mm, cross-section up to more than 120 μm.
GH 2 often with bright zone in shield. Leporidae
Key 4

b. Medulla of GH 1 and GH 2 without this characteristic
multicellular pattern with rows. 7

7a. Medulla of GH 1 and GH 2 filled in entire hair (Fig. 33). In
medullar slide, cortex always very thin, hardly or not visible.
Generally, cross-sections more than 150 μm in diameter. GH 1
coarse and wavy (Fig. 60), often collapsed; length *c.* 30–90 mm.
Except for the tip, the hair has approximately the same diameter
over its entire length. Artiodactyla
Key 12

b. Medullar pattern otherwise. 8

8a. GH 1 with black tip, white middle portion, and dark basal part.
Hair long: *c.* 50–100 mm. 9

b. GH 1 uniformly coloured or with different pattern. 10

9a. GH 1 with black tip (*c.* 15 mm), white middle portion, and dark
basal part; length up to *c.* 100 mm.
Shaft of GH 1 and GH 2 with transverse cuticular scales (Fig. 171).
Cross-section of GH 1 blunt oval, *c.* 125–150 μm, with wide
medulla (Fig. 102, compare with Fig. 101 = *Procyon lotor*),
medullar margins rather straight in thickest part of shield (Fig. 35).

57 *Nyctereutes procyonoides*
Key 9

b. Colour distribution of GH 1 almost identical, length *c.* 50–60 mm.
Cuticular pattern of GH 1 and GH 2 in shaft variable (Figs 68, 69,
72, or resembling Figs 70 and 71), scale margins either smooth or
rippled. Cross-section of GH 1 oblong, *c.* 125–150 μm, with
narrow medulla (Fig. 101), medullar margins rather straight in
thickest part of shield (Fig. 97). 58 *Procyon lotor*
Key 9

10a. Cuticula of GH 2 in shaft diamond petal (Figs 61, 62, and 63).
Transitional pattern between diamond petal pattern and simple
transverse scales in outermost base of shaft is short, i.e., less than
c. 100 μm (Fig. 64). Cuticula of GH 1 has variable structure.
Medulla of GH 1 and GH 2 often dark, their margins scalloped in
thickest part of shield with rounded bulges into cortex, visible
where oil has not penetrated (Figs 37, 270, 274). Medulla in shield
never isolated (Figs 25 and 31), often with opaque, transverse
oblong shapes tapering slightly at the ends and bordered by thin,
black lines. A cloisonné medullar pattern can usually be seen in the
shield if oil has penetrated (Figs 29 and 66, right sides).　Mustelidae
Key 10

b. Cuticula of GH 2 in shaft usually not diamond petal. If so, then
transitional pattern at least 300 μm long (Fig. 65a,b) or medullar
margins in thickest part of shield straight or fringed (Figs 35 and
36), to be seen if oil has not penetrated. Medullar pattern may be
isolated (Figs 25 and 31) or mustelid-like.　11

11a. Medulla of GH 1 and GH 2 in shield isolated, unicellular or
multicellular (Figs 25, 31, 88, 89, 92, 119, 121, 136, 147) or crescent
shaped (Figs 32, 112, 116). Medullar margins in thickest part of
shield may have scalloped appearance (Figs 119 and 147).　12

b. Medullar margins in thickest part of shield rather straight or
fringed (Figs 35 and 36), to be seen if oil has not penetrated. Cells
in shield never isolated or crescent shaped even after oil
penetration. GH 1 usually more than 25 mm long. If medullar
margins of GH 1 in thickest part of shield show some incisions, the
effect resembles an enlarged ladder pattern (Fig. 107; oil
penetration), the cells occupying the entire width of the medulla.
Width of hair considerable, up to *c.* 180 μm, GH 2 narrower with
narrow intermediate medulla (Fig. 28).　19

12a. Medulla of GH 1 in thickest part of shield shows brown more or
less crescent cells (Figs 32 and 116) if oil has penetrated. Cells often
contiguous and overlapping sometimes with similar shapes in the
brighter intercellular spaces (Figs 112 and 116). A cloisonné
pattern may be present. Cuticular pattern in proximal shaft
elongate petal (scales *c.* 6 × 30 μm, Fig. 113). In some parts of shaft
a diamond petal pattern may be present, but only over short
distances. In distal shaft and proximal shield, scales more
elongated and undulating, up to 80 μm long or longer in some
species, some of the scales V-shaped, tapering proximally (Figs 218
and 215, respectively). Cross-sections in thickest part of shield
c. 60–90 μm, oval, concavo-convex or with shallow concavities, in
distal shield never triconcave or quadriconcave, as shown in
Figs 230a–c, 231a, 233a,c. GH 1 *c.* 15–20 mm long.　Sciuridae
Key 5

b. No such crescent cells in medulla in thickest part of shield.
Cuticular scales in distal shaft and proximal shield less elongated.　13

13a. Medullar pattern of GH 1 in shield one cell layer (ladder) or two cells thick (Figs 121 (oil penetration), 119, 247). Occasionally a third cell layer is present. Cells usually almost opposing each other (Fig. 121). Medulla sometimes interrupted at constriction between shield and shaft. GH 1 and GH 2 less than 15 mm long, medulla in shaft ladder. Cross-sections *c.* 40–55 μm, circular, oval or concavo-convex (Fig. 122a–d), cuticula in shaft petal or diamond petal (Figs 123–133).   14

b.   Medulla of GH 1 in thickest part of shield three to more than seven cells thick. Usually they are clearly visible – even if oil has not penetrated – as dark isolated cells with almost equal length and width, sometimes they have an oblong appearance (Figs 25, 31, 136, 137, 140, 147). Although the cells may be more or less contiguous, a cloisonné pattern usually does not develop.   15

14a.   Cross-sections of GH 1 and GH 2 in shield mainly concavo-convex (Fig. 122d), cuticula in central shaft acute diamond petal or elongate petal in a regular way (Fig. 139). GH 1 up to *c.* 12 mm long, medulla in shield as in Fig. 119).   37 *Micromys minutus*
    Key 7

b.   Cross-sections of GH 1 and GH 2 mainly circular and oval (Figs 122a–c), cuticula in central shaft variable (Figs 123–125), less acute and less regular diamond or elongate petal as shown in Fig. 139.   Gliridae
    Key 8

15a.   Cuticular scale pattern of GH 1 and GH 2 in central shaft variable: diamond petal (Figs 138 and 141), acute elongate petal (Fig. 139), or more or less mosaic (Figs 135, 143, 144, 150). In proximal shield a similar arrangement (resembling Figs 126, 135, 165); scale margins at most slightly undulating, never wavy as in Fig. 239. Cross-sections in shield show 1, 2, or 3 concave sides (Figs 241–244), about 50–120 μm.   Subfamily: Murinae
    Key 7

> *Note*
> If cuticular scale pattern in shaft is diamond petal, then transitional petal pattern between simple outermost basal scales and diamond petal pattern is at least 300 μm long (Figs 65a,b) and not at most 100 μm as in Mustelidae (Fig. 64).

b.   Cuticular scale pattern in shaft and proximal shield otherwise.   16

16a.   Cuticular scale pattern of GH 1 and GH 2 in central shaft and especially in proximal shield longitudinal with undulating margins (Figs 155a,b, 156, 157 shaft, 239 proximal shield). Cross-sections oval or with 1–4 concave sides somewhere in the shield (Figs 230, 233, 236a,b, 237a,b), largest are *c.* 50–80 μm.   Subfamily: Cricetinae, Microtinae
    Key 6

b.   Cuticular scale pattern of GH 1 and GH 2 in shaft transverse (Figs 93, 158, 161).   17

17a.   Medulla of GH 1 in thickest part of shield narrow: as wide (Fig. 88) as or sometimes about half width of combined cuticula–

cortex, four to five cells wide, in central and distal shaft medulla as wide as or narrower than combined cuticula–cortex (Fig. 86), in basal part wider to much wider (Fig. 87). Medulla core alternatingly thicker and narrower over the entire hair. Length of GH 1 *c.* 50 mm, cross-sections oblong, up to 190 μm with small medulla (Fig. 90). Cuticula of GH 1 and GH 2 in shaft composed of narrow scales lying transversely (Fig. 91). No medulla present at boundary between shaft and shield in GH 2. UH medulla interrupted. 47 *Castor fiber*

b. Medulla of GH 1 in thickest part of shield wider than combined cuticula–cortex. Medulla of GH 2 present at boundary between shaft and shield. Medulla of UH not interrupted. 18

18a. Medulla of GH 1 in thickest part of shield less than three times width of combined cuticula–cortex (Fig. 162) e.g. CC/TW = 25/ 110 μm. Length of GH 1 *c.* 30 mm, cross-sections oblong, *c.* 100–120 μm (Fig. 160). 32 *Ondatra zibethicus*
Key 6

b. Medulla of GH 1 in shield wide, approximately four or five times width of combined cuticula–cortex (Fig. 89). CC/TW = 18–22/ 150 μm. Thickest part of shield about seven cells thick (Fig. 92). Cells in shaft somewhat longer and sometimes touching, forming more or less transverse streaks separated by light spaces at least equally wide. Cross-sections oblong, about 150 μm with long, narrow medulla (Fig. 95), cuticula of shaft with transverse scales (Fig. 93).
Length of GH 1 about 35–40 mm. GH 2 shorter, often with yellow band 5–6 mm long about 2 mm from tip. UH wavy, medulla with ladder pattern, not interrupted in basal half. 48 *Myocastor coypus*

19a. Medullar cells of GH 1 in thickest part of shield form a characteristic light–dark enlarged ladder pattern (visible if oil has penetrated). The bright parts are large and take up the entire width of the medulla (Fig. 107). Basal part of medulla resembles Fig. 106. Length of GH 1 *c.* 25–30 mm, cross-sections approximately circular over entire length, at widest part about 180 μm (Fig. 103), cuticula in basal part as in Figs 108 and 109. GH 2 curved, some of the cross-sections with one rather flat side appearing 'triangular' with rounded angles *c.* 110 μm (Figs 104 and 105), basal cuticula as in Fig. 110. UH elastic, shaft with zigzag, cross-sections predominantly oblong, with small medulla (Fig. 111), *c.* 40μm.
1 *Erinaceus europaeus* (hair on the belly)
Key 2

b. Medulla of GH 1 in thickest part of shield otherwise. 20

20a. Medullar margins in thickest part of shield fringed, visible if oil has not penetrated (Fig. 36). If oil has penetrated, outermost layer of shield shows a fine granular reversed cloisonné structure (Figs 30, 58, 168). Large and small agglomerations of black, grit-like particles can be seen in shield if oil penetration is partial (Fig. 177). Small light cells often present in the central layers of the medulla;

they can be almost circular as well as spool-shaped (Fig. 169, oil penetration).

Length of GH 1 30–60 mm, width of cuticula–cortex *c*. 15 μm, cross-sections circular to oval, some of the smaller ones more or less triangular (Fig. 172).

Cross-sections of GH 2 oval to oblong (Fig. 173), sometimes triangular, sections 1–4 almost circular. In some hairs basal shaft may be thicker than shield.

Felidae
Key 11

b. Medullar margins in thickest part of shield straight; visible if oil has not penetrated (Fig. 35). If oil has penetrated, a cloisonné pattern can often be seen distinctly in the outermost layer of the medulla (Figs 29, 174, 257). Length of GH 1 usually more than 30 mm, in some species up to 100 mm, width of cuticula–cortex up to 25 μm (sometimes smaller), cross-sections circular to oval, 90–140 μm (Figs 102, 167, 175).

Canidae
Key 9

Key 2

Insectivora: Erinaceidae, Soricidae, and Talpidae

Erinaceidae

Erinaceus europaeus differs strongly from the other insectivora owing to its dorsal spines and ventral hair. At a magnification of 40, the spines look like a maize cob, tapering close to the base. A typical cross-section taken at the thickest part of a spine is shown in Fig. 179. The hemispherical root implantation is very characteristic (Fig. 180). The species has been considered in key 1.

Soricidae and Talpidae

GH 1 is straight, GH 2 has a distinct zigzag shaft and a distinctly enlarged central shield (Fig. 56a and b). The cuticular scales along the shaft of GH 2 protrude in an alternate pattern (Fig. 56a), characteristic for these two families. GH 1 and GH 2 are less than 10 mm long. According to Vogel & Köpchen (1978), the cross-sections of straight hairs (Leithaare, corresponding with GH 0 and GH 1) of the genera Sorex and Neomys are circular. This only holds for GH 0, since GH 1 often shows complex and deep incisions (Fig. 56c).

1a. Cross-sections of GH 2 more or less H-shaped (Fig. 56d and e). 3

b. Cross-sections of GH 2 not H-shaped but more or less as in Fig. 56f and g or 181. 2

2a. Shaft of GH 2 with 5–7 zigzags, cross-sections oval (Fig. 181), roughly 40 × 20 μm. 9 *Talpa europaea*

b. Shaft of GH 2 with 3–5 zigzags, cross-sections more or less as in Fig. 56f and g. *Crocidura* 4

Key 2

3a. Cuticula of GH 2 at thickest part of shield and further distally with longitudinal connections between the transverse scales (Fig. 182). Longest scales at second zigzag of the shaft, proximally from the shield, less than 50 μm long (procedure for measuring shown in Fig. 184).   *Sorex*   6

 b. Cuticula of GH 2 at thickest and more distal parts of shield with less distinct longitudinal connections between transverse scales (Fig. 183). Longest scales at second zigzag more than 50 μm long.
                                                              Neomys   7

4a. Medulla of GH 1 and GH 2 unicellular and irregular (Figs 24 and 185). GH 1 often shows more than 30 connected air chambers forming letter-like symbols, but occasionally groups of only two to four solitary cells can be seen. Medulla of GH 1 in cross-sections often divided into two parts (Fig. 186).   8 *Crocidura leucodon*

 b. Medulla of GH 1 and GH 2 in shield unicellular and usually regular (Fig. 23). GH 1 often shows 10 to 15 solitary cells between letter-like configurations. GH 1 medulla in cross-sections sometimes divided into two parts.   5

5a. In GH 2, approximately 15 medullar cells per 100 μm. Maximum width of medullar cells of GH 2 in shield *c.* 25 μm. Shield of GH 2 often longer than, or as long as, shaft, cuticular pattern in shield with regular transverse scales and smooth margins (Fig. 187).
                                                           6 *Crocidura russula*

 b. In GH 2, 12 to 13 medullar cells per 100 μm in thickest part of shield. Maximum width of GH 2 medullar cells in shield *c.* 22 μm. Shield of GH 2 often shorter than shaft. Cuticula of GH 2 resembling *C. russula*.   7 *Crocidura suaveolens*

6a. Length of GH 1 and GH 2 *c.* 6 mm. Width of largest medullar cells in shield of GH 2 *c.* 27 μm. Cross-sections of GH 1 vary from circular (Fig. 188a) to H-shaped (Fig. 188b). Sides of cross-sections 2 and 3 of GH 2 in distal part of shield more or less parallel (Fig. 189a).   2 *Sorex araneus*

 b. Length of GH 1 and GH 2 4–5 mm. Width of largest medullar cells in shield of GH 2 *c.* 23 μm. Cross-sections of GH 1 variable but often H-shaped. Sides of distal cross-sections 2 and 3 of GH 2 more or less butterfly-shaped (Fig. 189b).   3 *Sorex minutus*

7a. Width of largest medullar cells of GH 2 in shield generally less than 18.5 μm. Cross-sections of GH 2 *c.* 28 μm wide. Cross-sections of GH 1 vary from circular to the shape in Fig. 190a, its medulla often divided into two parts.   4 *Neomys fodiens*

 b. Width of largest medullar cells of GH 2 in shield usually more than 18.5 μm. Cross-sections of GH 2 *c.* 33μm wide. Cross-section 4 of GH 1 usually shows only a minute incision (Fig. 190b).
                                                           5 *Neomys anomalus*

Key 3 Chiroptera: Rhinolophidae and Vespertilionidae

All representatives of Chiroptera in the geographical area dealt with in this study belong to two families, Rhinolophidae and Vespertilionidae. Unlike the other groups treated in this study, their hairs lack a medulla although they do have pigment (Fig. 191a and b). Therefore the category *Medulla* in the atlas section has been replaced by the heading *General impression*. The slides shown in photographs were prepared in the same way as the medullar material.

The length of GH 1 and GH 2 is less than 15 mm. GH 2 usually has a wavy or zigzag shaft with scales as in Fig. 191a (pigmentation may be otherwise). Usually GH 1 is slightly shorter than GH 2, slightly curved and often scarce, the single cuticular scales in the shaft do not protrude (Fig. 191b).

1a. Cuticular tips of GH 1 and GH 2 simple (Fig. 193a and b) with smooth margins. Cuticula of GH 1 and GH 2 in shield as shown in Fig. 192b. Shaft of GH 1 partly as shown in Fig. 192a, partly like Fig. 192b, in GH 2 entirely or partly as in Fig. 192a; normal and reversed K-shapes distinctly recognizable. Shaft of GH 2 transparent, without pigment concentrations, shown in Fig. 191a (use general impression slide). Cross-sections of GH 1 and GH 2 in thickest part of shield triangular (Fig. 194).   Rhinolophidae

> *Note*
> Distinction between the two species is difficult.
> The thickness of the hairs gives an indication.
>
> | | |
> |---|---|
> | Cross-section GH 1 *c.* 20.5 μm | 10 *Rhinolophus* |
> | GH 2 *c.* 17.5 μm | *ferrumequinum* |
> | GH 1 *c.* 17 μm | 11 *Rhinolophus* |
> | GH 2 *c.* 15 μm | *hipposideros* |

b. Cuticular structure in shaft mostly without normal and reversed-K shapes; if so cells at tip irregular as in Fig. 195.   Vespertilionidae   2

2a. Extreme tip of GH 2 (0.75 mm) has characteristic cuticular pattern shown in Fig. 195. A similar structure extends more proximally (Fig. 196). Cuticular pattern in shaft of GH 1 and GH 2 as shown in Fig. 197a and b, respectively. Both hair types transparent, without pigment concentrations in shaft as shown in Fig. 191a. Cross-sections of GH 1 and GH 2 circular to short-oval, widest diameter *c.* 23 and *c.* 17 μm, respectively.   22 *Nyctalus noctula*

b. Otherwise   3

3a. Cuticular scales in shield of GH 1 and GH 2 frilled (Figs 19, 198a, 200b).   4

b. Cuticular scales in shield of GH 1 and GH 2 without frills.   8

4a. Frills on GH 1 and GH 2 not very conspicuous, approximately

1 μm wide all over the shield and without any dentation
(Fig. 201a).
Shaft of GH 1 and GH 2 with pigment distributed more or less
evenly over the scales. Distal part of GH 2 shaft bears distinctly
protruding cuticular scales (Fig. 201b).   5

b. Frills in thickest part of shield of GH 1 and GH 2 conspicuous,
especially in distal part where some more than 2 μm, with a more
or less distinct dentation.   6

5a. Many of the largest cross-sections of GH 1 and GH 2 oval as
shown in Fig. 203a. Diameter of GH 1 then often more than 15.5 μm.
Length of GH 2 6–7 mm.   25 *Vespertilio murinus*

b. The largest cross-sections of GH 1 and GH 2 more or less circular
(see Fig. 203b). Diameter of GH 1 usually less than 15.5 μm.
Length of GH 2 7–8 mm.   26 *Barbastella barbastellus*

6a. Frills of GH 1 and GH 2 in thickest part of shield with distinct fine
dentation (Fig. 198a), especially in distal part. Structure of distal
150 μm of shield of GH 1 and GH 2 as shown in Fig. 198b
and c.   21 *Pipistrellus nathusii*

b. Frills of GH 1 and GH 2 only occasionally dentated (Figs 199a,
200b). Distal part may show dentation.   7

7a. Cuticular tip of GH 2 resembles that in Fig. 202b. Cuticular scale
pattern in distal part of the hair often resembles a zigzag line
(Fig. 199a).   20 *Pipistrellus pipistrellus*

b. Cuticular tip of GH 1 as shown in Fig. 200a, that of GH 2 as in
Fig. 202a. General impression of cuticular scale pattern very
complicated, a picture to which the frills contribute. Thickest part
of shield resembles Fig. 200b. Zigzag lines as mentioned under 7a
do not occur frequently.   23 *Nyctalus leisleri*

8a. Shaft of GH 2 undulating, sometimes with more than seven turns.
GH 2 in shield often shows a part with light tones (use
stereomicroscope and white background). Under these conditions,
a characteristic small dark tip can be seen. The hair seems to be
somewhat thinner in the light-toned part.
Cuticular of GH 1 and GH 2 in central shield appear as shown in
Fig. 204a. Length of GH 2 more than 10 mm, shield about 4 mm.
                                                              Plecotus spec.

> *Note*
> There are only slight differences between the two species of *Plecotus*.
> The cross-sections are similar in size (*c.* 18 mm), and also differ little
> as to shape. The light/dark contrast is stronger and more distinctive
> in the shaft of *P. austriacus* (nr. 28) than in that of *P. auritus* (nr. 27).

b. Shaft of GH 2 less undulating; hairs generally lack dark tip.   9

> *Note*
> *Myotis bechsteinii* (16) and *Vespertilio murinus* (25) sometimes have
> a small dark tip. *V. murinus* can be recognized instantly by its
> shorter hair (6–7 mm). Especially in *Myotis bechsteinii*, cross-
> sections nos 2–5 of GH 1 and GH 2 are shorter oval than those in
> Plecotus (Fig. 204c and b, respectively).

9a.   Cuticula of GH 1 and GH 2 in thickest part of shield with smooth
      as well as strongly rippled scale-margins (Fig. 205a), which is
      rather rare for bat hairs. The rippled pattern decreases toward the
      tip. The cuticula sometimes shows slight indications of a frill.
      Cross-sections of GH 1 and GH 2 appear as in Fig. 205b and c,
      respectively.

                                                    24 *Eptesicus serotinus*

  b.   Scale margins smooth. Shaft of GH 2 usually show strong pigment
       concentrations in distal part of each scale (Fig. 191a).

                                                                Myotis spec.

       > *Note*
       > Differences between *Myotis* species are slight. Some indications are
       > given in the following tentative key. More material and more
       > elaboration is needed to validate these distinctions.

  I   Length of GH 2 less than 9 mm   14 *M. emarginatus*
                                                          15 *M. nattereri*
                                                          18 *M. daubentonii*
                                                          19 *M. dasycneme*
      GH 2 9–10 mm   13 *M. brandtii*
      GH 2 longer than 10 mm   II
  II Cross-sections of GH 2 *c.* 16 μm in diameter, very bright and
      showing scattered pigment granules (Fig. 204c). Extreme tip of
      hair often dark (use stereomicroscope and white background).
                                                          16 *M. bechsteinii*
      Cross-sections of GH 2 *c.* 17–19 μm. Cuticula of GH 2 in thickest
      part of shield often with more than 40 scales within 300 μm. Last
      300 μm of shield more than 60 scales.   12 *M. mystacinus*
      Cross-sections of GH 2 *c.* 21 μm bright and showing scattered
      pigment granules.   17 *M. myotis*

Key 4   Lagomorpha: Leporidae

The family of the Leporidae is one of the easiest to recognize because of its
specific medulla. In the shield of GH 1 and GH 2 this medulla is composed
of approximately five to eight rows of cells arranged in a regular pattern
(Fig. 26). In the shaft there are three or four of these rows.

   The shaft of GH 1 and GH 2 consists of narrow longitudinal cuticular
scales up to 90 μm long and sometimes even longer, which may run somewhat
obliquely (Fig. 206). Alternating with this type of pattern, especially in the
proximal part of the shaft, there may be shorter cells with undulating margins
(Fig. 207). The distal part of the shaft and usually the proximal part of the
shield as well show a pattern like that in Fig. 208. The cuticula in the central

part of the shield shows regular waves (Fig. 13) as well as irregular wave patterns (Fig. 14), in the middle part, usually with smooth margins and laterally with smooth or rippled margins. A longitudinal groove is present on either side of the shield of GH 1 and sometimes on that of GH 2, causing the dumb-bell shape in cross-sections.

Hair length is 25–35 mm, cross-sections are *c.* 125–150 μm wide. Shields of GH 2 usually have a yellow zone.

1a. Cross-sections of GH 1 and GH 2 in central and distal parts of shield dumb-bell shaped (Fig. 212).
Basal half of pigmented hair grey, seen especially in bundles of hair.   54 *Oryctolagus cuniculus*

  b. Cross-sections of GH 1 dumb-bell shaped, especially in black hairs (do not confuse with the longer black GH 0 with long tapering tip). Cross-sections of GH 2 in thickest part of shield concavo-convex (Figs 210, 211), sometimes roughly dumb-bell shaped in proximal and distal parts of shield (Fig. 213). Basal half of hair white (seen in bundles).   52 *Lepus europaeus*
                                                    53 *Lepus timidus*

> *Notes*
> 1. Keller (1980) distinguishes between *L. europaeus* and *L. timidus* by the shapes of the medullar cells in cross-sections. In *L. europaeus* they should be rounded, in *L. timidus* angular. However, we did not see this feature in our specimens.
> 2. The shafts of hairs on the head and legs of *Lepus* may be dark as in *Oryctolagus*.

Key 5   Rodentia I: Sciuridae

The Sciuridae are related to the Muridae with respect to the medulla, especially in the proximal shield; the pattern of the isolated (dark) cells is hardly distinguishable from that in the Muridae. In the thickest part of the shield and more distally the resemblance is less pronounced. In this part, the brown medullar cells form a crescent pattern (Figs 32, 112, 116).

Cuticula of GH 1 and sometimes GH 2 in the distal shaft of *Sciurus vulgaris* and *S. carolinensis*, but less often in those of *Tamias sibiricus*, show undulating scale margins closely resembling those of Leporidae at that place (Figs 114b, 215). In GH 1, this pattern often extends into the proximal shield.

1a. Length of GH 1 *c.* 15 mm, cross-sections up to 60–70 μm wide, in distal half of shield approximately circular (Fig. 115b), in proximal part slightly oval, sometimes asymmetrical and/or biconcave (as in Fig. 115d). Cross-sections of GH 1 and GH 2 in thickest part of shield usually short oval (Fig. 220a and b). Proximal and central parts of shaft of GH 1 and GH 2 show broad petal cuticular pattern (Fig. 216). Fragments of V-shaped scales may be found in distal part of shaft and proximal part of shield of GH 1 (Fig. 117a). Cuticular shaft of GH 2 resembles that in Fig. 118. GH 2 shorter, often with a lighter zone in shield.   51 *Tamias sibiricus*

b. Length of GH 1 *c.* 20 mm, cross-sections more than 70 μm wide. Cross-sections of GH 1 and GH 2 variable. Proximal and central parts of shaft with narrow longitudinal scales (Figs 113, 114a, 218), distal shaft and proximal shield of GH 1 usually show V-shaped scales with undulating margins (Fig. 218).   *Sciurus*   2

2a. Cross-sections of GH 1 and GH 2 in thickest part of shield concavo-convex (Figs 115a, 217). Cross-sections *c.* 85 μm wide.

49 Sciurus vulgaris

b. Cross-sections of GH 1 and GH 2 circular to oval (Figs 115b,c,e, 219, 220a,b), with at most shallow concavities in GH 2 (Fig. 115d). Cross-sections 70–80 μm wide.   *50 Sciurus carolinensis*

Key 6

Rodentia II: Muridae; Cricetinae and Microtinae

In Muridae the medullar cells can usually be seen as dark, isolated cells in the shield even in the absence of oil (Fig. 31). These are almost as long as they are wide, sometimes longer (Figs 136, 137, 140, 221). Although they may touch each other, the pattern remains one of isolated cells (Fig. 31) rather than a cloisonné pattern (Fig. 29).

In the thickest part of the shield of GH 1 the multicellular, medullar pattern may be up to six cells wide (Figs 136, 221), that of GH 2 usually with one or more cells less. In the shaft the pattern is usually of the ladder type (resembling that in Figs 224, 225, 226).

GH 1 is slightly curved, the shield distinctly thickened as in GH 2 (Fig. 229). In most representatives one of the sides of the shield is concave, with a cuticular impression resembling those in Figs 209 and 222.

Identification may lead to the assignment of the species *Rattus rattus* and *Mus musculus* (Murinae) to this key, at the basis of some similarity between features. However, in *Mus musculus* the cuticular margins in the shaft and proximal shield are much straighter (Figs 148–151) than in Microtinae, where they are wavy. Both species are dealt with in this key.

1a. Some of distal cross-sections of GH 1 and GH 2 are concave on all four sides (Fig. 230a and b). Larger cross-sections in the central part of shield as in Fig. 230c, *c.* 55–60 μm. Medulla of GH 1 usually four cells wide. Length of GH 1 *c.* 14 mm.

30 Clethrionomys glareolus

b. Cross-sections of distal GH 1 and GH 2 show fewer than four concave sides.

2a. Cross-sections 2 and 3 either bell-shaped (Figs 45 and 231a) or concavo-convex (Figs 42 and 47). Larger cross-sections as in Fig. 233b or d (*c.* 60–85 μm). Cuticular scales of GH 1 and GH 2 in distal half of shaft and proximal shield as in Fig. 155a and b (shaft), resembling those in Figs 207 and 239 (proximal shield),

never with rather straight margins as in Figs 148–51. Medulla of
GH 1 usually four cells wide. Length of GH 1 *c.* 12 mm.

> 33 *Pitymys subterraneus*
> 34 *Microtus arvalis*
> 35 *M. agrestis*
> 36 *M. oeconomus*

Note
It is difficult to distinguish between these four species. *M. oeconomus*
is the only species that generally has many pigment granules in the
cortex of the basal part of the shaft. These occur especially where
the medulla starts and proximally from it (Fig. 232).

b. Distal cross-sections never bell-shaped.   3

3a. Cuticular margins of central shaft and proximal shield always
    rather straight, unlike the Microtinae; pattern dependent on side
    of hair used for slide; see Fig. 149 (concave side) and Fig. 150
    (convex side). Concave side of shaft and proximal shield highly
    specific (Figs 149 and 151). The patterns in central and distal parts
    of shield (concave side) resemble those in Figs 152 and 153,
    respectively. Medulla of GH 1 four cells wide. In shaft of GH 1 and
    GH 2 medulla usually predominantly multicellular. Cross-sections
    of GH 1 and GH 2 concavo-convex in thickest part of shield
    (Fig. 154), never biconcave as in Fig. 233d, *c.* 55–60 μm. Length of
    GH 1 *c.* 10 mm.   43 *Mus musculus* (Murinae)

b. Cuticular scale pattern in shaft and proximal shield (concave side)
    otherwise. Cross-sections much larger.   4

4a. Cuticular scales of GH 1 and GH 2 transverse in central shaft
    (Fig. 158). Cross-sections of GH 1 100–120 μm (Fig. 160), medulla
    in thickest part of shield less than three times larger than the
    cuticula–cortex (e.g. CC/TW = 25/110 μm, Fig. 162). Length of
    GH 1 *c.* 30 mm.   32 *Ondatra zibethicus*

b. Scales longitudinal in central shaft (Figs 238, 227, 228).   5

5a. Largest cross-sections of GH 1 concavo-convex in thickest part of
    shield *c.* 80–100 μm or more (Fig. 145), distal cross-section nr. 3 of
    GH 1 distinctly more elongated oval than that in Fig. 237a.
    Cuticular scale pattern in shaft of GH 1 and GH 2 undulating as
    in Figs 156, 157, 159. Medulla of GH 1 usually five or six cells wide.
    Length of GH 1 and GH 2 *c.* 20 mm.   42 *Rattus rattus* (Murinae)

> *Note*
> Brunner & Coman (1974) reported cross-sections up to 240 μm, but
> we have not found such large hairs among our specimens.

b. Distal cross-sections (1, 2, and 3) of GH 1 circular to short oval
    (Figs 236a, 237a, cross-section 3). Diameter in thickest part of
    shield at most *c.* 80 μm. Medulla of GH 1 usually not more than
    five cells wide.   6

6a. Largest cross-section of GH 1 as in Fig. 236b, *c*. 60–70 μm. Margins of cuticular scales of GH 1 wavy as in Figs 238 (central shaft), 239, and 240 (proximal part of shield). Medulla of GH 1 as a rule four to five cells wide. Length of GH 1 15–16 mm.

31 *Arvicola terrestris*

b. Largest cross-section of GH 1 as in Fig. 237b, *c*. 80 μm. Margins of cuticular scales of GH 1 less wavy, as in Figs 227 (central shaft), 234, and 235 (proximal part of shield). Medulla of GH 1 usually five cells wide. Length of GH 1 *c*. 20 mm.

29 *Cricetus cricetus*

Key 7   Rodentia III: Muridae; Murinae

The medulla in Murinae shows the same isolated structure as in Cricetinae/ Microtinae. In GH 1 it may be up to seven cells wide in the thickest part of the shield, in GH 2 the number is usually one or more cells smaller.

In most of the representatives, cuticular patterns of GH 1 and GH 2 in the shaft approximate the diamond petal type (Figs 133, 138, 139, 141). However, *Mus musculus* (Figs 149, 150) and *Rattus rattus* (Fig. 143) show some variability and the pattern may even resemble that of the Microtinae. Such cases of these species have already been treated in key 6.

The transitional pattern intermediate between the transversely arranged cuticular scales in the outermost proximal shaft and the diamond petal pattern is of the petal type (Fig. 65a and b). It expands over a distance of at least 300 μm long (Fig. 65a), whereas in the Mustelidae this segment is at most *c*. 100 μm long (Fig. 64). Cross-sections of some species strongly resemble those of the Microtinae and are of approximately the same size (Figs 44, 241b–d) others are oval or concavo-convex.

1a. Cross-sections of GH 1 and GH 2 large, 80–120 μm or more in diameter, never with concavities on three sides as in Fig. 244a or shaped as in Fig. 244d and e anywhere in distal part of shield. Length of GH 1 *c*. 15–25 mm.   2

b. Cross-sections of GH 1 and GH 2 mostly smaller, less than 80 μm. If not, cross-sections in distal part of shield shaped as described in 1a. Length of GH 1 *c*. 12 mm.   3

2a. Cross-sections of GH 1 in thickest part of shield concavo-convex with a rather deep concavity (Fig.145). In GH 2 this character is less pronounced. Cuticula of GH 1 and GH 2 in shaft irregular diamond petal (Figs 135, 144), in its wider part and proximal shield as in Fig. 245. Medullar pattern of isolated type, five to six cells wide in thickest part of shield.   42 *Rattus rattus*

b. Cross-sections of GH 1 in thickest part of shield oblong with sometimes one or both sides slightly concave (Fig. 146). Cuticula in central shaft of GH 1 diamond petal (Fig. 138), in proximal part as in Fig. 246. Medullar pattern of isolated type, six to seven cells wide in GH 1 (Fig. 147) and five to six cells in GH 2.

41 *Rattus norvegicus*

1. Brunner & Coman (1974) reported cross-sections up to 200 μm and 240 μm in the brown and black rat, respectively. However, we have not found such large hairs among our specimens.

2. Occasionally GH 0 is circular in cross-section (*c.* 70 μm). This type of hair is slender with a less flattened shield, its medulla is not isolated, but closely resembles that of the Mustelidae in the thickest part of the shield; however, it is very narrow: about 1.5 times the width of the combined cuticula–cortex, while it is at least four times that in Mustelidae.

3a. Distal cross-sections of GH 1 and GH 2 circular to oval. In thickest part of shield concavo-convex (Figs 154, 243).   4

b. Always some distal sections of GH 1 and GH 2 with concavities on two or three sides (Figs 241b–d, 244a).   5

4a. Cuticular scale pattern of central part of shaft dependent on side of hair used for the slide, e.g. Fig. 149 (concave side) and Fig. 150 (convex side). Concave side of shaft and proximal shield highly specific (Figs 149, 151). The patterns resemble those in Figs 152 and 153 in central and distal part of shield (concave side) respectively. Medulla in shield of GH 1 four cells wide, in shaft of GH 1 and GH 2 usually predominantly multicellular, cross-sections 55–60 μm (Fig. 154).   43 *Mus musculus*

b. Cuticula of GH 1 and GH 2 in central shaft elongate petal to diamond petal (Figs 132, 133, 139). Medulla in thickest part of shield of GH 1 at most three cells wide (Fig. 119), but often unicellular and if so the pattern is usually of the letter type. In shaft of GH 1 and GH 2 medulla unicellular. Cross-sections 45–50 μm (Fig. 122d).   37 *Micromys minutus*

5a. Cross-sections 3, 4, 5, and 6 of GH 1 and GH 2 distinctly concave on three sides (Fig. 241). Diameter *c.* 55–75 μm.   38 *Apodemus sylvaticus*
     39 *Apodemus flavicollis*

b. Distal cross-sections of GH 1 have only shallow concavities (Figs 244d,e) especially on the short sides (Fig. 244a). Sections of proximal part of shield usually concavo-convex (Fig. 244b,c). Medulla five to six cells wide. Cross-sections of GH 1 *c.* 70–80 μm
     40 *Apodemus agrarius*

Key 8   Rodentia IV: Gliridae

The medullar shaft and often the shield of GH 1 and GH 2 of the Gliridae is of the ladder type (Figs 120, 134). The shield, however, may have two layers of cells in which dark and light cells sometimes alternate. This feature shows up especially in parts of the preparation into which oil has penetrated. The medulla usually shows a letter-type pattern in GH 1 (Figs 24, 247). The cuticula of GH 1 and GH 2 is rather variable, as illustrated by Figs 123–131 (shaft) and 248–253 (central part of shield). Hair length up to *c*. 15 mm, cross-sections circular to oval (Fig. 122a–c) and at most 55 μm wide.

1a. Medulla of GH 1 often interrupted at transition between shaft and shield. Combined cuticula–cortex approximately as wide as medulla in thickest part of shield (Fig. 121). In GH 2, which is zigzag, cuticula–cortex also wide. Cuticular pattern intermediate, as shown in Fig. 129 (shaft) and Figs 252 and 253 (shield). Cross-section oval, medulla relatively small and indistinct (Fig. 122a), diameter about 55 μm GH 1 straight to slightly curved. **44 *Glis glis***

b. Medulla of GH 1 not interrupted at transition between shaft and shield. Cuticular pattern in shaft more longitudinal (Figs 123, 125, 128, 130, 131). Combined cuticula–cortex in shield distinctly smaller than medulla. Both GH 1 and GH 2 usually show cuticular ladder pattern in shield. **2**

2a. Cuticular scales in shaft of GH 1 all distinctly longer than wide (Figs 123, 125, 128), in proximal part of shaft GH 2 of length and width sometimes equal (Fig. 127). Cuticular patterns of GH 1 resemble those in Figs 126 and 248 in proximal shield, and Fig. 250 in central shield. Central shield of GH 2 regular wave (Fig. 249), as is part just proximally from central shield. Cross-sections of GH 1 oval, *c*. 40 μm (Fig. 122b). **45 *Muscardinus avellanarius***

b. Some cuticular scales in the shaft of GH 1 and GH 2 as long as wide (Figs 130, 131). Cuticular pattern of GH 1 in central part of shield regular wave (Fig. 251), approximately the same as in *M. avellanarius* (Fig. 126). Proximal and central shield of GH 2 regular wave, also resembling *M. avellanarius*. Cross-sections of GH 1 oval, diameter *c*. 35 μm (Fig. 122c), in GH 2 approximately oblong, *c*. 40 μm (Fig. 254). **46 *Eliomys quercinus***

Key 9   Carnivora I: Canidae and Procyonidae

Canidae have a cloisonné medulla that can be seen well after oil penetration. Where the medulla is somewhat narrower, the cloisonné structure is the same but its appearance is different due to the tapering of the medulla (Fig. 255).

The medulla is very dark if oil has not penetrated. At best, all that can be seen under these conditions is a somewhat rounded, transverse pattern with some lighter parts (Fig. 256). The same type of medulla occurs in Mustelidae. However, this family can be readily distinguished from the Canidae by two distinct features. First, the Mustelid GH 2 shaft always shows a diamond petal cuticula (Figs 62 and 63), whereas in Canidae the pattern is normally transverse and petal (Figs 69–71, 73, 75, 170, 171); and, second, in Canidae the medullar margins in the thickest part of the shield are rather straight (Figs 35 and 256), lacking the more or less rounded bulges into the cortex occurring in Mustelidae (Fig. 37).

The Procyonid *Procyon lotor* has been included in this key because of its close relationship with the Canidae. Its medulla resembles that of the Canidae in the absence of oil, but a cloisonné pattern is either absent or obscure.

Vulpes vulpes sometimes has a cuticular pattern in the shaft that closely resembles the diamond petal pattern of Mustelidae. In such cases this species might be assigned to the family of the Canidae via the mustelid key. On the other hand, the mustelid *Meles meles* may end up in the canid key because of its rather straight medullar margins in the thickest part of the shield. The latter species is included in this key.

Few differences are found between the dog and the red fox where hairs of the same length are compared, but there are so many breeds of dogs, that the dimensions and shape of their hairs can vary widely and this renders identification difficult. Cross-sections are magnified 200 instead of 400 times.

1a. Cross-sections of GH 1 in thickest part of shield usually more than 125 μm. Hairs often show black or white areas in distal shield. 2

 b. Cross-sections of GH 1 in thickest part of shield less than 125 μm. Colour of hair variable. 4

2a. Largest cross-sections of GH 1 show wide medulla, length/width ratio of the latter approximately 2:1 (Fig. 102).
57 *Nyctereutes procyonoides*
(further information is given below)

 b. Largest cross-sections of GH 1 have narrow medulla. Ratio length/width of the latter approximately 3:1 (Figs 101 and 157). 3

3a. Long hair: GH 1 *c.* 90 mm, GH 2 *c.* 45 mm long. Tip and base of GH 1 and GH 2 often white, middle part black or dark. Smaller type of GH 2 with specific irregular pattern in proximal shaft, which can be reduced to a diamond petal pattern (Fig. 59). Medullar structure in distal half of hair is fine-grained, and circular and oval cells can be seen vaguely if oil has penetrated (Fig. 267 = GH 1). Cross-sections of GH 1 oval, up to 200 μm.
65 *Meles meles*
(Mustelidae)
(For further information see Mustelidae key 10: 3b)

 b. GH 1 *c.* 50–60 mm long, sometimes more. Tip and base of GH 1 and GH 2 often dark, middle light or white. Cuticula of GH 1 and GH 2 sometimes divergent (Fig. 68 or similar to those in Figs 69–71), never as in Fig. 59. Medulla of GH 1 in shield approximately as in Fig. 97 (oil penetration).
Cross-sections of GH 1 *c.* 125–150 μm. 58 *Procyon lotor*
(further information is given below)

4a. Especially the pointed ends of medullar cells of GH 1 and GH 2 along the borderline between the medulla and cortex in shield expressive and striking (Fig. 166). Cuticular shaft of GH 1 and GH 2 transverse (Figs 73 and 75) or longitudinal (Figs 74 and 76), in GH 2 often petal to broad diamond petal (Fig. 77). 55 *Vulpes vulpes*
(further information is given below)

b.   Pointed ends missing or greatly reduced. Cuticula of GH 1 and
      GH 2 in shaft show more or less mosaic pattern (Fig. 170).

                                                           56 *Canis familiaris*
                                          (further information is given below)

55 *Vulpes vulpes*

Length of GH 1 and GH 2 up to *c.* 60 mm and *c.* 45 mm,
respectively; red brown, usually with a variably light zone in shield.
Medulla in the shaft often with pointed cells along the borderline
between medulla and cortex (Fig. 262), but sometimes like that in
Fig. 261. Cloisonné pattern in outer medullar layer very distinct
in proximal shield after oil penetration. Shrunken cells forming the
cloisonné borders around the light parts are rather coarse-grained.
These borders become wider in thickest part of shield, finally
changing into a reversed cloisonné pattern of coarse-grained cells
with narrow hyaline borders (Fig. 30 or resembling Fig. 168).
Combined cuticula–cortex *c.* 9–12 μm there. In shield, cuticula
usually rippled (Fig. 263).
Cross-sections of GH 1 and GH 2 *c.* 90–120 μm. GH 1 rather
circular, GH 2 short-oval (Fig. 167).

56 *Canis familiaris*

Length of GH 1 and GH 2 dependent on breed, up to *c.* 100 mm.
Cross-sections very close to those of *Vulpes vulpes*, in GH 1 almost
round, in GH 2 oval (Fig. 175). Medulla in shield normally very
opaque, oil hardly penetrating. Combined cuticula–cortex *c.* 20 μm
or more. Cloisonné borders in shield strongly pigmented (Fig. 264,
proximal shield). Cuticular patterns at curves in the shaft of GH 2
resemble those in Fig. 265. Cuticular scale margins in shield rippled
(Fig. 73 proximal shield, Fig. 263 central and distal shields).

57 *Nyctereutes procyonoides*

Long hairs sometimes more than 100 mm and often show *c.* 15 mm
black tip, more proximally an almost white segment of variable
length, basal part dark. Medullar pattern cloisonné (Fig. 257),
especially in proximal part of hair, visible if oil has penetrated
(Fig. 255). Somewhat deeper in medulla, rounded cells visible
(Fig. 258 = proximal shield). Width of combined cuticula–cortex in
thickest part of shield *c.* 20 μm or more.
      Shaft of GH 1 and GH 2 with transverse cuticular scales
(Fig. 171). In central part of hair, scales rippled (Fig. 259). Cuticula
in shield may vary strongly, some of the hairs regular to irregular
wave with very narrow (3–5 μm) and elongated transverse scales
(slightly rippled) in middle part, laterally often heavily rippled. In
other hairs pattern resembles that in Fig. 260 or rippled as in
Figs 99 and 263. Cross-sections blunt/oval (125–180 μm). Distal
sections 3 and 4 may be either oval or circular.

58 *Procyon lotor*

Length of GH 1 *c.* 50–60 mm, often with dark tip and almost white
field in shield approximately like that of *N. procyonoides*. Cross-
sections of GH 1 oblong, *c.* 125–150 μm, sometimes broad, light

cuticular border. Medulla of GH 1 in proximal shaft with transverse, more or less spool-shaped structures (Figs 94 and 96, with and without oil penetration, respectively), in shield approximately as in Fig. 97 (after oil penetration). Outer medullar layer, although sometimes obscure, may show cloisonné pattern in shaft and shield (Fig. 29 right). Large and small, variably dense concentrations of dark almost round air bubbles can be seen after partial oil penetration. Combined cuticula–cortex in thickest part of shield rather broad (*c.* 25–30 μm). Cuticula of GH 1 and GH 2 in shaft may vary strongly (shaft: Fig. 68 or similar to Figs 69–71, its scale margins either smooth or rippled; shield: Figs 69, 72, 98 or totally rippled as in Fig. 99), areas proximally from tip sometimes very characteristic (Fig. 100).

Key 10   Carnivora II: Mustelidae

The family of Mustelidae is characterized by the diamond petal cuticula of GH 2 (Figs 61–63, 76, 165). The medullar margins of GH 1 and GH 2 are often dark and in the thickest part of the shield show rounded bulges protruding into the cortex (Figs 37, 274) as opposed to the fine structure and straight margins of the Canidae and Felidae (Figs 35, 36, 256, 257). This feature can be observed best in the absence of oil. The only exception is *Meles meles*, which has straight medullar margins.

In many mustelids the pattern of the opaque medulla in the shield is characterized by opaque transverse, oblong shapes tapering slightly at the ends and bordered by thin black meandering lines. After oil has penetrated a well-defined cloisonné pattern is visible (Figs 66 and 163 shield, Fig. 164 shaft) except in *Lutra lutra*, *Mustela vison*, and *Meles meles*. The width of the combined cuticula–cortex in the thickest part of the shield is often *c.* 12 μm and at most 25 μm except in *Meles meles* and *Lutra lutra*, where it may be even greater. The length of the transitional cuticular pattern between the outermost basal transverse scales and the diamond petal scales is not great, at most 100 μm (Fig. 64), and is shorter than in Murinae, where it is *c.* 300 μm or more (Fig. 65a,b). Cuticula of GH 1 in the shaft may vary strongly: sometimes diamond petal as GH 2, but usually completely different and resembling the patterns shown in Figs 70, 73, 75, 78, 79 or 80–82.

Cross-sections of GH 1 and GH 2 vary from *c.* 90–150 μm. The hair profiles of GH 1 and GH 2 are very similar within individuals; the smaller ones are considered to be GH 2 (see e.g. species no. 64, Part III (atlas).

Length of GH 1 and GH 2 varies enormously (*c.* 8–90 mm) within Mustelidae. *Mustela vison*, *Meles meles*, and *Lutra lutra* may differ in some of the features mentioned above.

Because the canid *Vulpes vulpes* may erroneously be assigned to this mustelid key, it is also included. Illustrations of cross-sections are magnified 200 instead of 400 times.

1a. Medulla of GH 1 shows, just proximally from the thickest part of shield, very dark and narrow cells *c.* 3–6 μm wide and *c.* 9–15 μm long, transversely arranged against cortex between lighter-toned narrow spaces *c.* 3 μm wide. Many concentrations of dark pigment present in this part of shield (Fig. 85). Medulla of GH 1 and GH 2 in thickest part of shield resembles that in Fig. 163. The diamond petal scales seem to be separated in shaft; this effect is due to the position of the proximal margins of the scales, which are out of focus when the central and distal parts of the scales are in focus (Fig. 61). Cross-sections of GH 1 and GH 2 oblong, *c.* 155 μm (Fig. 84). Length of GH 1 *c.* 18–28 mm.   62 *Mustela vison*

b. Medullar pattern otherwise.   2

2a. Medullar margins rather straight in thickest part of shield (if oil has not penetrated). No rounded bulges protruding into the cortex as in Figs 37 and 67; at most, some fibril-like structures can be seen (Figs 35 and 256).   3

b. Medulla in thickest part of shield shows rounded bulges protruding into the cortex if oil has not penetrated (Figs 37 and 67).   4

3a. Length of GH 1 and GH 2 up to *c* 60 and *c.* 45 mm, respectively. Both types often reddish-brown, shield somewhat lighter or with light zone. Cross-sections of GH 1 rather round, diameter *c.* 90–120 μm. GH 2 oval and with same diameter (Fig. 167). Cuticula of GH 2 and often GH 1 in shaft as in Figs 74 and 76. Medulla of GH 1 and GH 2 in thickest part of shield often has pointed ends along the borderline between medulla and cortex (Fig. 166, oil penetration), in shaft as in Fig. 262 (oil penetration) but may also be as in Fig. 261. Cloisonné pattern in outer medullar layer very distinct in proximal shield if oil has penetrated. More distally in shield a reversed cloisonné pattern of coarse-grained cells with narrow hyaline borders can be seen (as in Fig. 30 or resembling Fig. 168).   55 *Vulpes vulpes* (Canidae)
Key 9

b. Length of GH 1 and GH 2 up to *c.* 90 and *c.* 45 mm, respectively. Both types often show remarkable white tip, proximally from that the hair is black, the remaining basal part light or white. Cross-sections of GH 1 oblong: up to more than 200 μm wide (Fig. 57). GH 2 also oblong, smaller: *c.* 130 μm. Combined cuticula–cortex of GH 1 wide: CC/TW = *c.* 45 μm/210 μm. In shaft, dimensions approximately the same.
Medullar structure in distal half of hair fine-grained and showing vaguely circular and oval cells after oil has penetrated (Fig. 267 = GH 1). After partial penetration of oil, scattered variably sized agglomerations of black, grit-shaped, particles become visible (Figs 58 and 177). Cuticular pattern of GH 1 in shaft as well as in larger type of GH 2 variable (Fig. 266 or as in Figs 68 and 69). Cuticular scale margins in shield often rippled (Fig. 18), sometimes as in Fig. 268, especially in proximal part of shield. The smaller type of GH 2 show a specific irregular pattern in the

proximal shaft, but this can sometimes be reduced to a diamond petal pattern (Fig. 59). 65 *Meles meles*

> *Note*
> Parts of the cuticular pattern of GH 1 and GH 2 are often missing in gelatin slides. In such cases the pattern has a fine-grained longitudinally arranged structure. It seems possible that these parts of the hairs have been eroded by the burrowing behaviour characterizing this species.

4a. GH 1 and GH 2 less than 20 mm long, and as a rule less than *c.* 13 mm. Cuticula in shaft diamond petal in all GH 1 and GH 2 (Fig. 62). 5

 b. GH 1 and GH 2 more than 20 mm long, often even more than 30 mm. Cuticula of GH 2 in shaft diamond petal (Figs 62, 63, 165), GH 1 in shaft resembles that in Fig. 80–82 or 70, 73, 75, 78, or 79. However, some GH 1 may have a diamond petal cuticula in the shaft. 6

5a. Length of GH 1 generally less than 9 mm. Medullar margins in thickest part of shield show rather deep incisions as in Fig. 269 (to be seen if no oil has penetrated). Usually less than 40 bulges protruding into the cortex over 300 μm. Cross-sections of GH 1, long oval *c.* 110 μm long (Fig. 271a). 60·*Mustela nivalis*

> *Note*
> This and the following species are difficult to distinguish due to some overlapping of characters.

 b. Length of GH 1 generally 9–13 mm. Medullar margins in thickest part of shield have shallower incisions (Fig. 270). Usually 45–50 bulges protruding into the cortex over 300 μm. (Next to these cells the tops of somewhat deeper-lying cells are visible, blurring the image and hindering accurate counting.) Cross-sections of GH 1 oval, *c.* 110 μm (Fig. 271b). 59 *Mustela erminea*

6a. Length of GH 1 *c.* 20–25 mm. Cloisonné pattern does not occur. Where oil has penetrated large medullar cells are visible, occupying the complete width of the medulla (Fig. 83). If no oil has penetrated, these cells can hardly be seen; the medulla then resembles a poorly stirred assortment of fluids with different turbidity. Cuticula–cortex rather wide. CC/TW = 30–35 μm/ 145 μm. UH without medulla or with strongly interrupted medulla. Cuticular pattern of shield variable, proximally as in Figs 272 and 276, in the middle as in Fig. 276; distally, scale margins usually rippled (Fig. 18). Shaft of GH 2 and usually GH 1 diamond petal. Cross-sections of GH 1 and GH 2 oblong, up to 160 μm with long narrow medulla. 66 *Lutra lutra*

 b. Length of GH 1 *c.* 30–45 mm. If oil has penetrated, a distinct cloisonné pattern is visible (Figs 66, 163, right). If oil has not penetrated, medulla in shield rather dark, but often transverse cells are often visible, then bordered by thin meandering dark lines

(Figs 66, 163, left). Cuticula in shaft of GH 2 always diamond petal (Figs 63, 165), in GH 1 sometimes with a transverse pattern (Figs 80–82). Cross-sections of GH 1 and GH 2 oval, roughly $90 \times 120 \ \mu m$.

7

7a. GH 1 and GH 2 often heavily pigmented. Many dark pigment spots along the dark lines of the cloisonné pattern of medulla in shield after oil penetration (Fig. 163, right side). These spots obscure the cloisonné pattern somewhat. The irregularity of the pattern of medulla in distal shield is due to its relative density (Fig. 275). The smaller type of GH 2 generally shows a cloisonné pattern which is very transparent and distinct in the shield.

61 *Mustela putorius*

b. Only a few pigment spots present (Fig. 66, right)

8

8a. If no oil has penetrated, the dark transverse medullar cells often lie approximately perpendicular to the cortex of the central shield (Fig. 66, left side). Distally from this area the medulla shows the original character rather than a somewhat disordered pattern, as in Fig. 274.

63 *Martes martes*

b. Medullar cells in middle part of shield usually do not lie perpendicular to the cortex, but obliquely (Fig. 67; compare with Fig. 66, left side). Distally to this area the medullar pattern becomes somewhat disordered (Fig. 274); the cells become less distinct and shorter. This pattern is sometimes observed in thickest part of shield as well. If oil has penetrated, the inner medullar cells appear rather round in the part of the shield just distally from the middle part.

64 *Martes foina*

Key 11   Carnivora III : Felidae

The Felidae are separated from the other families because of the rather straight medullar margins with fibril-like fringes (Figs 36 and 58; the latter shows partial oil penetration). This feature invariably holds for the thickest part of the shield. The medulla in the rest of the hair may, however, show rounded bulges protruding into the cortex. If oil has penetrated, a typical fine-grained transparent structure is seen in the outer layer of the medulla of the middle part of the shield; this structure is crisscrossed by colourless and non-granular narrow lines (Fig. 58). These lines may form a reversed cloisonné pattern (Figs 30 and 168). More or less round or short-oval cells are often indistinctly visible in the inner medulla of the shield (Fig. 169), which is sometimes as large as the total width of the medulla locally. The cloisonné pattern is absent in the thickest part of the shield, but sometimes occurs in the shaft.

Large and small agglomerations of black grit-like particles can be seen in the shield if oil has only partially penetrated (Fig. 58). Medullas which are less heavily pigmented in the shield and into which oil has not penetrated are identical to mustelid medullas except for the margins (Figs 66, left side and 67).

The cuticula of GH 1 varies considerably in the shaft (Figs 70, 73, 75, 77, 78, 170, 171). The proximal shield shows a transverse scale pattern (Fig. 79 and/or as in Fig. 259), the central and distal shields are mostly rippled.

The shaft of GH 2 normally has broad diamond petal scales as in Fig. 78, sometimes alternating with the pattern in Fig. 70. The proximal shield of GH 2 resembles Fig. 71 and/or Fig. 79, sometimes via a pattern like that in Fig. 259.

Cross-sections of GH 1 are predominantly round to short-oval; some are more or less triangular (Fig. 172). Cross-sections 1–4 of GH 2 are approximately circular. Proximally they approach oblong (Fig. 173), and some are triangular. Cross-sections of GH 1 approximate $c.$ 90–120 μm in diameter. The length of GH 1 varies from $c.$ 30 to 60 mm. The basal shaft of GH 1 and GH 2 may be wider than the shield. Small differences are found between the wild and the domestic cat. The distinction given below is based chiefly on the dimensions, which are normally larger in wild cat. All cross-sections are magnified 200 instead of 400 times.

1a. Length of GH 1 up to $c.$ 40 mm, GH 2 $c.$ 30 mm. Cross-sections of GH 1 rather round to short oval (Fig. 172), GH 2 oval, both diameters approximately $c.$ 90–100 μm. **68 *Felis catus***

b. Length of GH 1 up to $c.$ 60 mm, GH 2 $c.$ 40 mm. GH 1 sometimes and GH 2 often show yellow zone in the shield. Cross-sections of GH 1 round to short oval, GH 2 oblong (Fig. 173), diameters of both approximately 120 μm. **67 *Felis silvestris***

> *Note*
> The medullar margins of *Meles meles* (Mustelidae) are also rather straight in the thickest part of the shield. If oil has not penetrated they closely resemble the Felidae (Fig. 36). However, *Meles meles* lacks the fribril-like fringes visible in Felidae if oil has not penetrated. Moreover, other features such as cross-sections, combined cuticula–cortex value, and hair length are completely different. *Procyon lotor* and *Nyctereutes procyonoides* (Procyonidae and Canidae, respectively) have straight medullar margins as well in the thickest part of the shield too. However, the combined cuticula–cortex of GH 1 is greater than $c.$ 15 μm ($c.$ 30 and 20–25 μm, respectively).

Key 12 Artiodactyla: Suidae, Cervidae, and Bovidae

The suid *Sus scrofa* is easily distinguished from all other groups by its coarse, thick, and exceedingly opaque GH 1, which is always split at the top (Fig. 279).

The representatives of the families Cervidae and Bovidae treated here, i.e. *Cervus dama*, *C. elaphus*, *Capreolus capreolus*, and *Ovis musimon*, can also be easily distinguished from all other groups by their specific medulla. The cells do not form regular longitudinal rows, as in Leporidae (Fig. 26), but four to six angular cells arranged irregularly fill the entire hair. The cortex, which is clearly present in all other orders, is paper thin and consistently seems to be absent (see illustrations of the medulla of species no. 70 in the atlas section and Fig. 33).

GH 1 and GH 2 are sometimes collapsed caused by desiccation and are identical with respect to the medulla and cuticula. After oil penetration, a very distinct and clear cloisonné pattern can be seen (Fig. 277).

The transverse cuticular pattern is of the mosaic type and is similar in all four of these species (Fig. 278).

Although predominantly specific, characteristics of these species sometimes overlap and this complicates correct identification. Some reserve must therefore be exercised in applying this key. For all *Artiodactyla*, cross-sections are magnified 100 times.

> *Note*
> 1. The thin wavy UH of Cervidae and Bovidae, which is approximately 12–15 μm thick measured in the medullar slide, has no medulla at all, as in Chiroptera, but is much longer than 15 mm, which makes it easy to distinguish between them.
> 2. Cross-sections, cuticula, and medulla, especially for long hairs, are illustrated in the atlas section, and are considered as GH 1.

1a.   GH 1 long, straight, and very coarse; split at top, often more than once (Fig. 279). The latter feature is present even in short hairs, where the split parts are only a few mm long. In long hairs (80–120 mm) the split parts may be up to 25 mm long and be split again.

Hairs have the same thickness all over and are up to 120 mm long, the diameter proximally from the split point can be up to 400 μm. Cuticula of GH 1 and GH 2 rippled all over the hair, ripples distally very close. Medulla of GH 2 in shaft interrupted and narrow, distinctly narrower than combined cuticula–cortex; in distal part as wide as or wider.   *69 Sus scrofa*

  b.   Hairs undulating, not split at top (Fig. 280).   2

2a.   Combined cuticula–cortex in cross-sections of central part of GH 1 rather thick, approximately 6–16 μm. Number of undulations in a hair about 40 mm long is generally less than twelve. Length and diameter of hair highly variable but generally substantial, length normally 40–60 mm (but up to 80 mm), diameter of cross-sections 200–400 μm. Distal cuticular scale margins vary from smooth to heavily rippled.   *71 Cervus elaphus*

> *Note*
> In some individuals of this species another type of hair occurs, here called GH 1A. This type is straight or slightly undulating, approximately 30 mm, and very flat. In cross-sections the ratio between the shortest and longest diameter is about 4:1 (e.g. 340×80 μm).

  b.   Combined cuticula–cortex in cross-sections usually at most 6 μm.   3

3a.   Undulations irregular, usually seven or fewer in a hair of *c.* 30 mm. Cross-sections vary considerably in diameter (150–400 μm). Combined cuticula–cortex in cross-sections of middle part of GH 1 3–6 μm thick. Cuticular scale margins usually rather smooth. Transverse scales in basalmost shaft rather short (Fig. 282).
       70 Cervus dama

b.   Undulations regular, approximately ten or more.   4

4a. Undulations considerable: from *c*. 12 to more than 20 according
     to hair lengths of 30 to 50 mm, respectively. Diameter
     approximately 200 μm. Combined cuticula–cortex in cross-sections
     of middle part of GH 1 *c*. 3–6 μm thick. Cuticula in basalmost part
     of shaft more or less reticular, as in Fig. 281 and unlike Fig. 282.
     Length of GH 1 30–50 mm.   73 *Ovis musimon*

b.   Approximately 10 to 13 undulations in a hair 35–40 mm long.
     Combined cuticula–cortex in cross-sections of middle part of GH 1
     up to 3 μm. Characteristic V-shaped incisions, usually in rows, in
     distal cuticular pattern (Fig. 283). Transverse cuticular scales in
     basalmost shaft somewhat longer than in *Ovis musimon* (Fig. 281)
     and *Cervus dama* Fig. 282). Cross-sections up to 350 μm, generally
     short-oval to oval (Fig. 284).   72 *Capreolus capreolus*

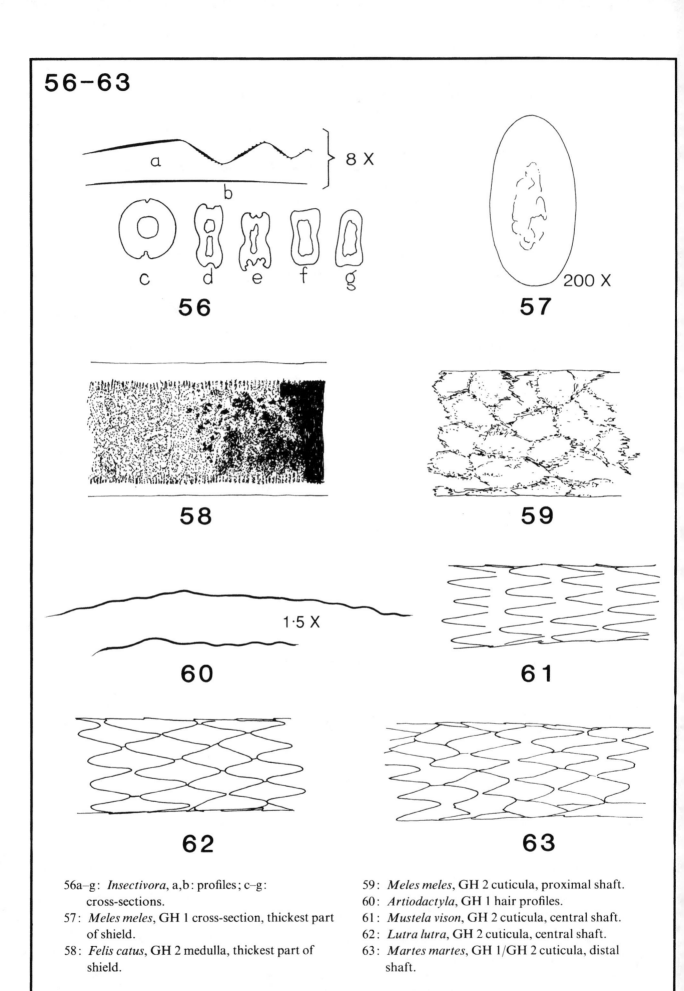

56

57

8 X

200 X

58

59

1·5 X

60

61

62

63

56a–g: *Insectivora*, a,b: profiles; c–g: cross-sections.

57: *Meles meles*, GH 1 cross-section, thickest part of shield.

58: *Felis catus*, GH 2 medulla, thickest part of shield.

59: *Meles meles*, GH 2 cuticula, proximal shaft.

60: *Artiodactyla*, GH 1 hair profiles.

61: *Mustela vison*, GH 2 cuticula, central shaft.

62: *Lutra lutra*, GH 2 cuticula, central shaft.

63: *Martes martes*, GH 1/GH 2 cuticula, distal shaft.

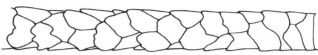

a

64

b **65**

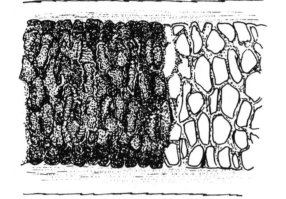

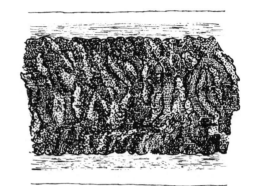

66

67

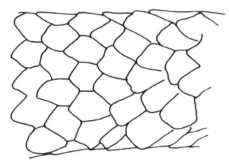

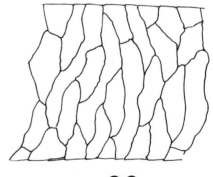

68

69

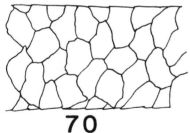

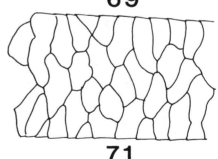

70

71

64: *Mustela erminea*, GH 1 cuticula, proximal shaft, transitional pattern.

65: *Rattus norvegicus*, GH 2 proximal shaft, transitional pattern; b: GH 1 proximal shaft, transitional pattern.

66: *Martes martes*, GH 1 medulla, thickest part of shield.

67: *Martes foina*, GH 1 medulla, central shield.

68: *Procyon lotor*, GH 2 cuticula, proximal shaft.

69: Idem, GH 2 cuticula, proximal shield.

70: *Felis domesticus*, GH 1 cuticula, proximal shaft.

71: Idem, GH 1 cuticula, proximal shaft.

72-80

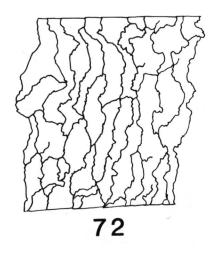

72

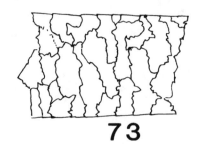

73

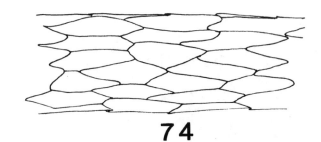

74

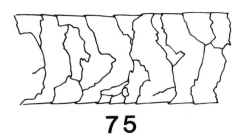

75

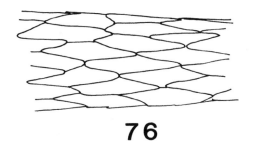

76

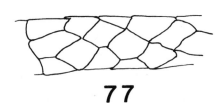

77

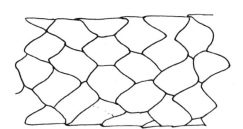

78

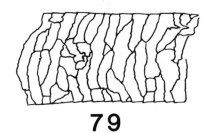

79

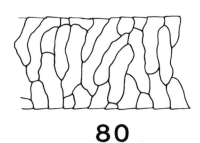

80

72: *Procyon lotor*, GH 1 cuticula, proximal shield.
73: *Canis familiaris*, GH 2 cuticula, proximal shield.
74: *Vulpes vulpes*, GH 1 cuticula, shaft.
75: Idem, GH 1 cuticula, central shaft.
76: Idem, GH 1 cuticula, shaft.

77: Idem, GH 2 cuticula, shaft.
78: *Felis catus*, GH 2 cuticula, shaft.
79: Idem, GH 1 cuticula, proximal shield.
80: *Martes martes*, GH 2 cuticula, proximal shield just distally to transitional part.

81–88

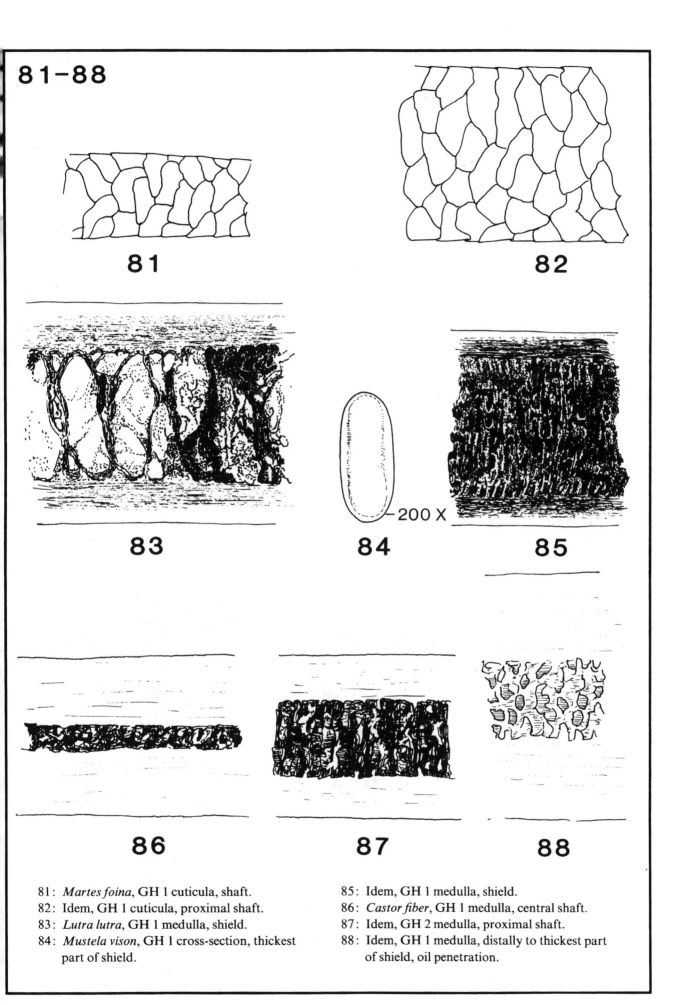

81: *Martes foina*, GH 1 cuticula, shaft.
82: Idem, GH 1 cuticula, proximal shaft.
83: *Lutra lutra*, GH 1 medulla, shield.
84: *Mustela vison*, GH 1 cross-section, thickest part of shield.

85: Idem, GH 1 medulla, shield.
86: *Castor fiber*, GH 1 medulla, central shaft.
87: Idem, GH 2 medulla, proximal shaft.
88: Idem, GH 1 medulla, distally to thickest part of shield, oil penetration.

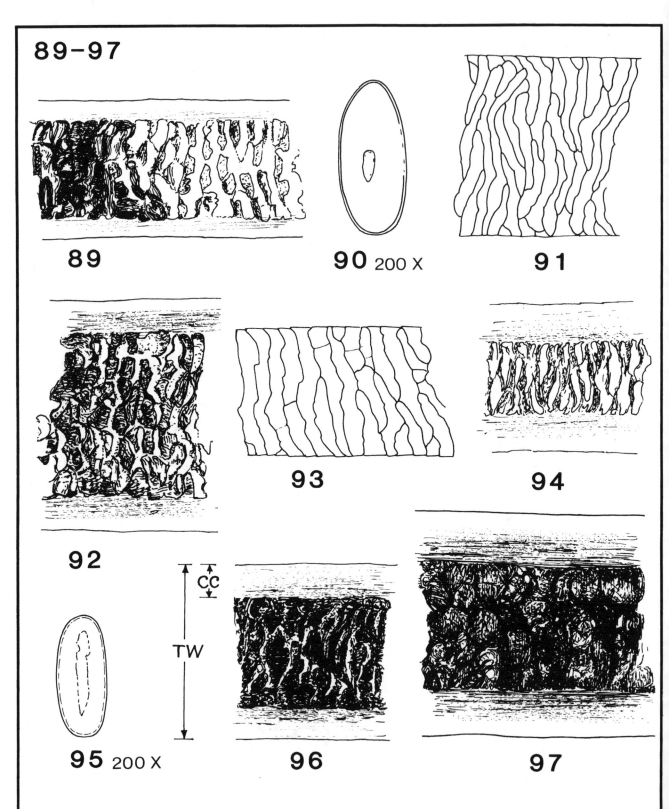

89 – 97

89

90 200 X

91

92

93

94

95 200 X

96

97

89: *Myocastor coypus*, GH 1 medulla, central shaft.
90: *Castor fiber*, GH 1 cross-section, thickest part of shield.
91: Idem, GH 1 cuticula, proximal and central shaft.
92: *Myocastor coypus*, GH 1 medulla, thickest part of shield.

93: Idem, GH 1 cuticula, central shaft.
94: *Procyon lotor*, GH 1 medulla, proximal shaft, oil penetration.
95: *Myocastor coypus*, GH 1 cross-section, thickest part of shield.
96: *Procyon lotor*, GH 1 medulla, shaft.
97: Idem, GH 1 medulla, shield.

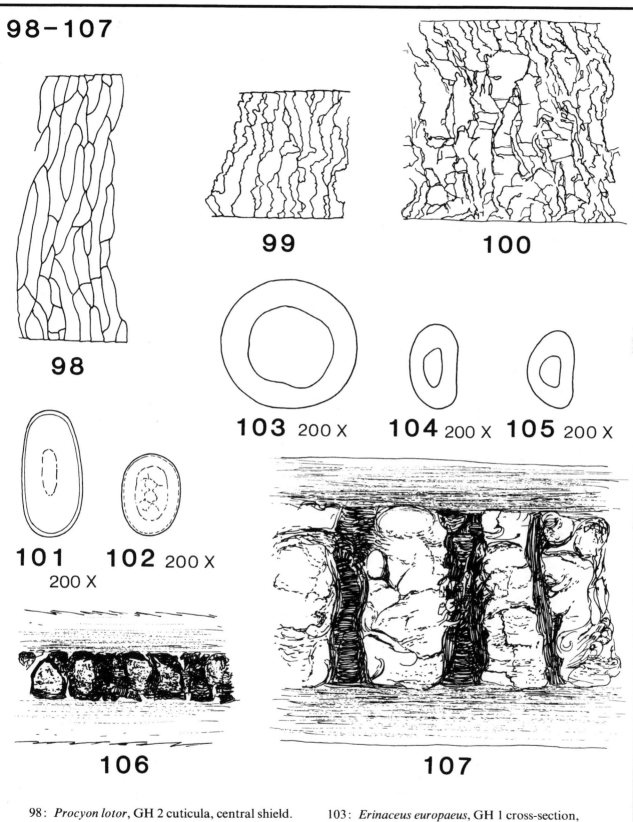

98

99

100

101 200 X **102** 200 X

103 200 X **104** 200 X **105** 200 X

106

107

98: *Procyon lotor*, GH 2 cuticula, central shield.
99: *Felis catus*, GH 1 cuticula, shield.
100: *Procyon lotor*, GH 1 cuticula, distal shield.
101: Idem, GH 1 cross-section, thickest part of shield.
102: *Nyctereutes procyonoides*, GH 1 cross-section, near thickest part of shield.

103: *Erinaceus europaeus*, GH 1 cross-section, thickest part of shield.
104: Idem, GH 2 cross-section, thickest part of shield.
105: Idem, GH 2 cross-section, proximal shield.
106: Idem, GH 1 medulla, basal part.
107: Idem, GH 1 medulla, thickest part of shield.

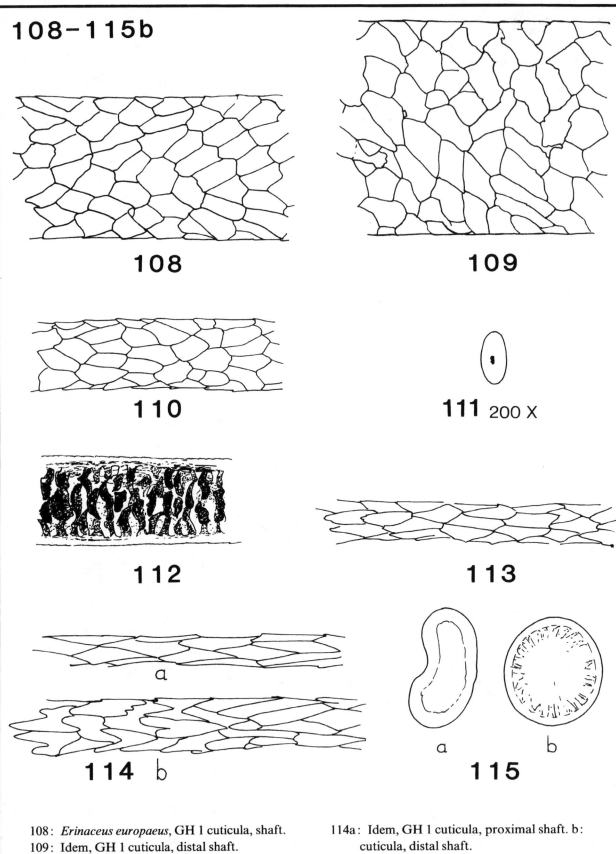

108 – 115b

108

109

110

111 200 X

112

113

114 b

115

108: *Erinaceus europaeus*, GH 1 cuticula, shaft.
109: Idem, GH 1 cuticula, distal shaft.
110: Idem, GH 2 cuticula, basal part.
111: Idem, UH cross-section, proximal shield.
112: *Sciurus vulgaris*, GH 1 medulla, central shaft.
113: Idem, GH 1 cuticula, proximal shaft.

114a: Idem, GH 1 cuticula, proximal shaft. b: cuticula, distal shaft.
115a: Idem, GH 1 cross-section, thickest part of shield. b: *Tamias sibiricus*, GH 1 cross-section 4.

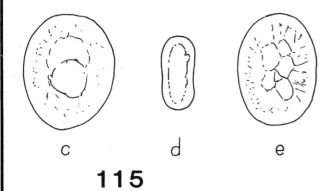

c   d   e

115

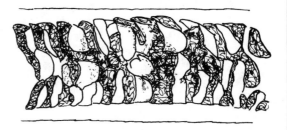

116

118

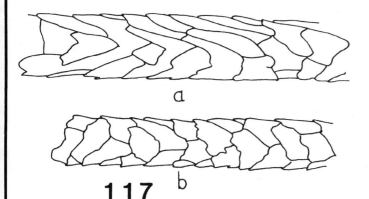

a

b

117

119

120

121

a   b   c   d

122

115c: *Tamias sibiricus*, GH 1 cross-section 5. d:
  Sciurus carolinensis, GH 2 cross-section,
  thickest part of shield. e: Idem, GH 1 cross-
  section, thickest part of shield.

116: *Tamias sibiricus*, GH 1 medulla, central
  shield.

117a: Idem, GH 1 cuticula, distal shaft. b: Idem,
  GH 1 cuticula, central shaft.

118: Idem, GH 2 cuticula, near central shaft.

119: *Micromys minutus*, GH 1 medulla, thickest
  part of shield.

120: *Glis glis*, GH 1 medulla, shaft.

121: Idem, GH 1 medulla, thickest part of shield,
  oil penetration.

122a: Idem, GH 1 cross-section 6. b: *Muscardinus
  avellanarius*, GH 1 cross-section 7. c: *Eliomys
  quercinus*, GH 1 cross-section 5. d: *Micromys
  minutus*, GH 1 cross-section 3.

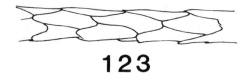

123

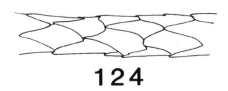

124

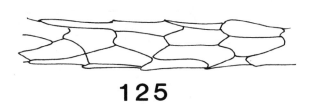

125

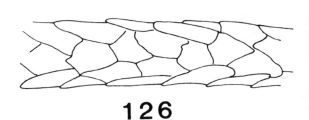

126

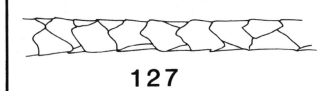

127

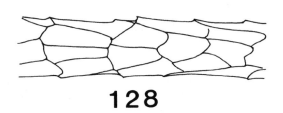

128

129

130

131

132

123: *Muscardinus avellanarius*, GH 1 cuticula, proximal shaft.
124: Idem, GH 1 cuticula, central shaft.
125: Idem, GH 1 cuticula, shaft.
126: Idem, GH 1 cuticula, distal shaft/proximal shield.
127: Idem, GH 2 cuticula, central shaft of second

zigzag before the shield.
128: Idem, GH 1 cuticula, distal shaft.
129: *Glis glis*, GH 1, GH 2 cuticula, shaft.
130: *Eliomys quercinus*, GH 1 cuticula, shaft.
131: Idem, GH 1 cuticula, distal shaft.
132: *Micromys minutus*, GH 1 cuticula, proximal shaft.

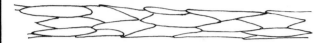

133

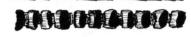

134

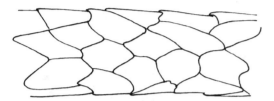

135

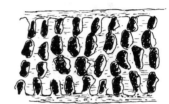

136

137

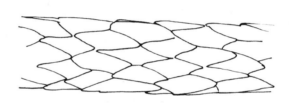

138

139

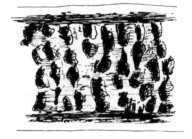

140

141

142

133: *Micromys minutus*, GH 1 cuticula, central shaft.
134: *Glis glis*, GH 1 medulla, shaft.
135: *Rattus rattus*, GH 1 cuticula, basal part.
136: *Clethrionomys glareolus*, GH 1 medulla, central shield.
137: Idem, GH 1 medulla, central shield.

138: *Rattus norvegicus*, GH 1 cuticula, central shaft.
139: *Micromys minutus*, GH 1 cuticula, shaft.
140: *Cricetus cricetus*, GH 1 medulla, thickest part of shield.
141: *Apodemus sylvaticus*, GH 2 cuticula, central shaft.
142: *Micromys minutus*, GH 1 medulla, shaft.

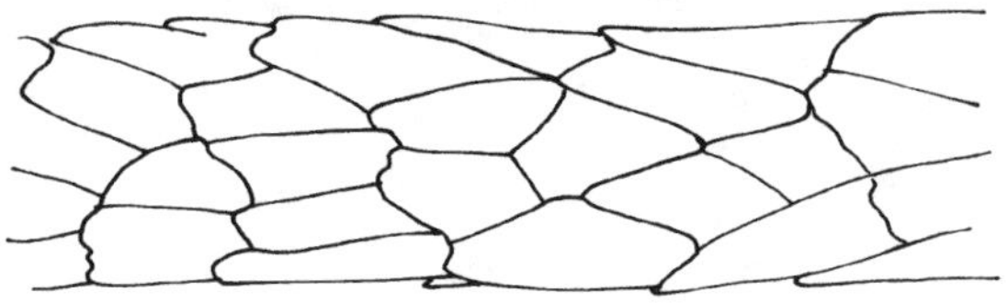

143

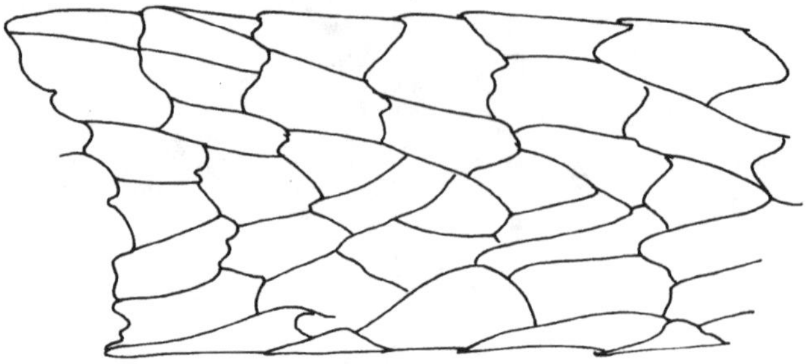

144

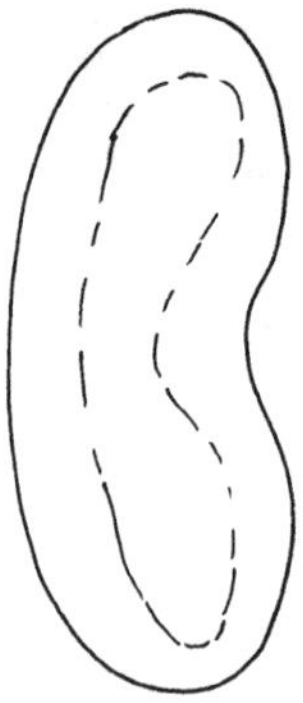

145

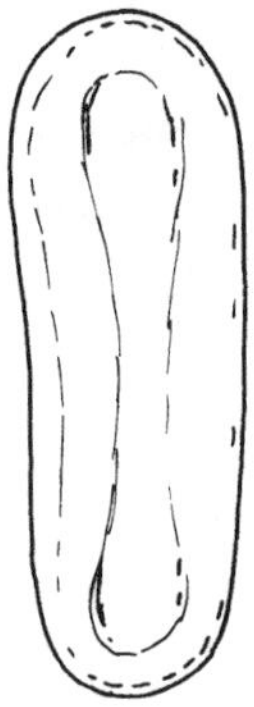

146

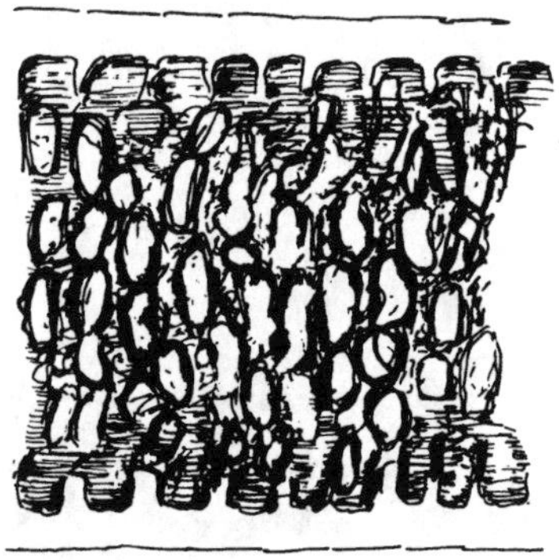

147

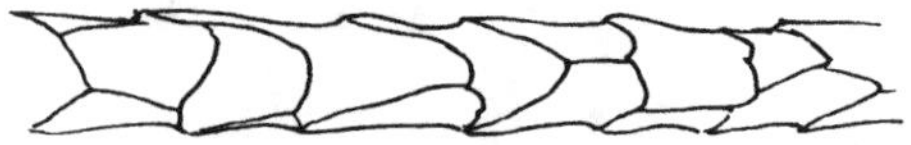

148

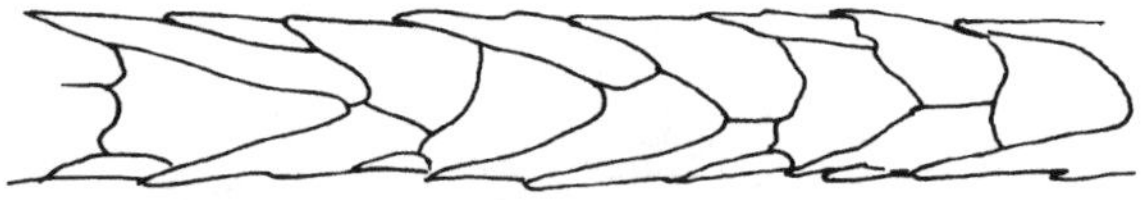

149

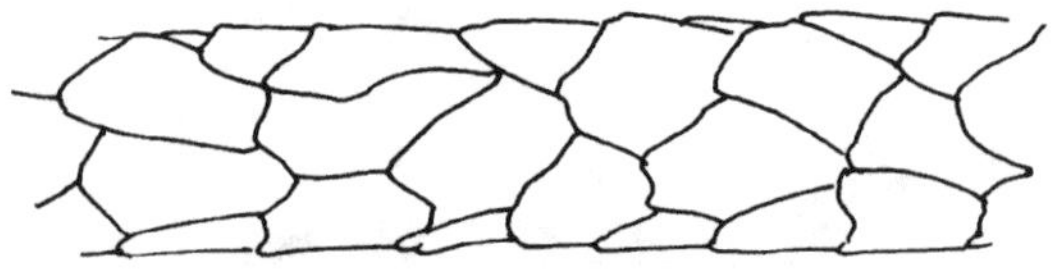

150

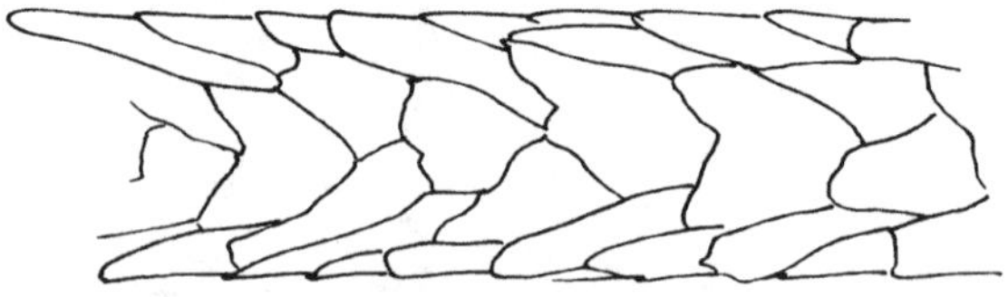

151

143: *Rattus rattus*, GH 2 cuticula, central shaft.
144: Idem, GH 1 cuticula, shaft, concave side.
145: Idem, GH 1 cross-section 5.
146: *Rattus norvegicus*, GH 1 cross-section 4.
147: Idem, GH 1 medulla, thickest part of shield.
148: *Mus musculus*, GH 1 cuticula, proximal shaft.

149: Idem, GH 1 cuticula, central shaft, concave side.
150: Idem, GH 2 cuticula, central shaft, convex side.
151: Idem, GH 1 cuticula, just proximal from the central shield, concave side.

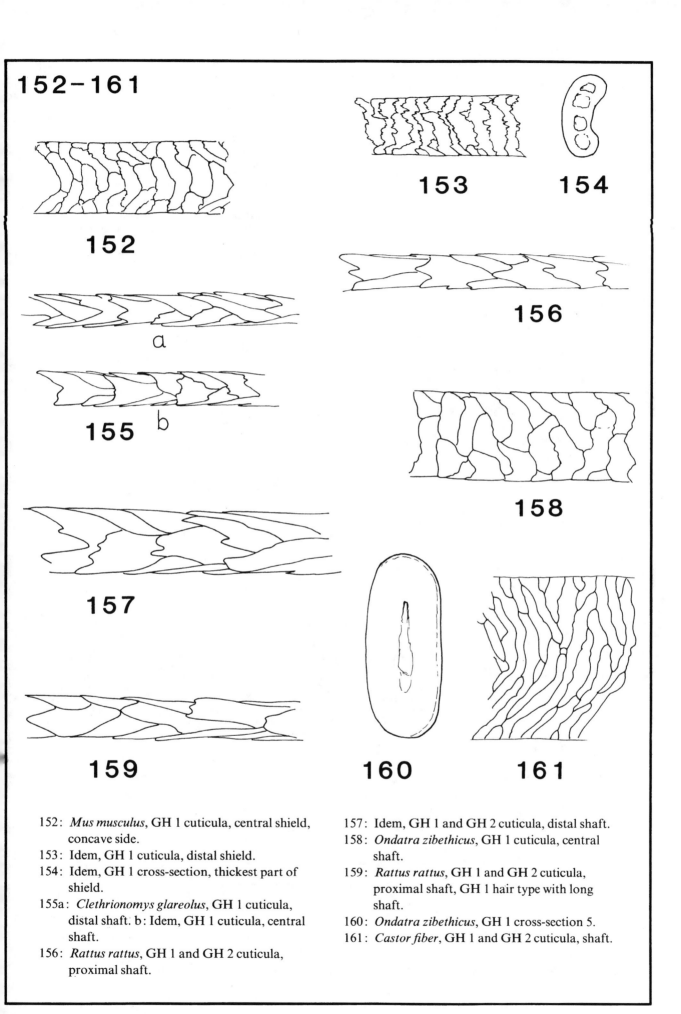

152: *Mus musculus*, GH 1 cuticula, central shield, concave side.

153: Idem, GH 1 cuticula, distal shield.

154: Idem, GH 1 cross-section, thickest part of shield.

155a: *Clethrionomys glareolus*, GH 1 cuticula, distal shaft. b: Idem, GH 1 cuticula, central shaft.

156: *Rattus rattus*, GH 1 and GH 2 cuticula, proximal shaft.

157: Idem, GH 1 and GH 2 cuticula, distal shaft.

158: *Ondatra zibethicus*, GH 1 cuticula, central shaft.

159: *Rattus rattus*, GH 1 and GH 2 cuticula, proximal shaft, GH 1 hair type with long shaft.

160: *Ondatra zibethicus*, GH 1 cross-section 5.

161: *Castor fiber*, GH 1 and GH 2 cuticula, shaft.

162

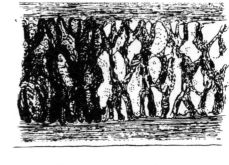

163

164

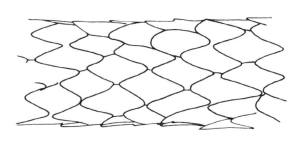

165

166

167 200 X

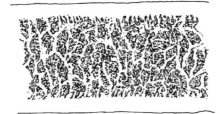

168

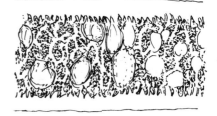

169

162: *Ondatra zibethicus*, GH 1 medulla, shield, oil penetration.

163: *Mustela putorius*, GH 2 medulla, central shield.

164: *Martes martes*, GH 1 medulla, shaft, oil penetration.

165: *Mustela putorius*, GH 1 cuticula, shaft.

166: *Vulpes vulpes*, GH 1 medulla, thickest part of shield.

167: Idem, GH 2 cross-section, thickest part of shield.

168. *Felis catus*, GH 1 medulla, central shield, oil penetration.

169: Idem, GH 2 medulla, shield, oil penetration.

170-177

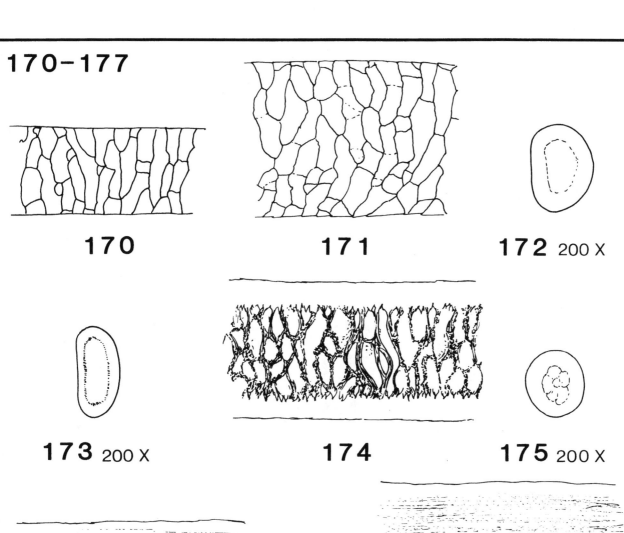

170

171

172 200 X

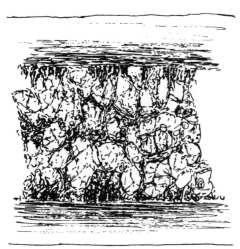

173 200 X

174

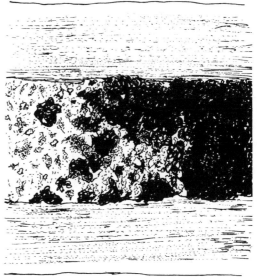

175 200 X

176

177

170: *Canis familiaris*, GH 2 cuticula, central shaft.
171: *Nyctereutes procyonoides*, GH 1 cuticula, proximal shaft.
172: *Felis catus*, GH 1 cross-section, thickest part of shield.
173: *Felis silvestris*, GH 2 cross-section, thickest part of shield.
174: *Meles meles*, GH 2 medulla, shield.

175: *Canis familiaris*, GH 2 cross-section, thickest part of shield.
176: *Procyon lotor*, GH 1 medulla, thickest part of shield.
177: *Meles meles*, GH 1 medulla, shield.

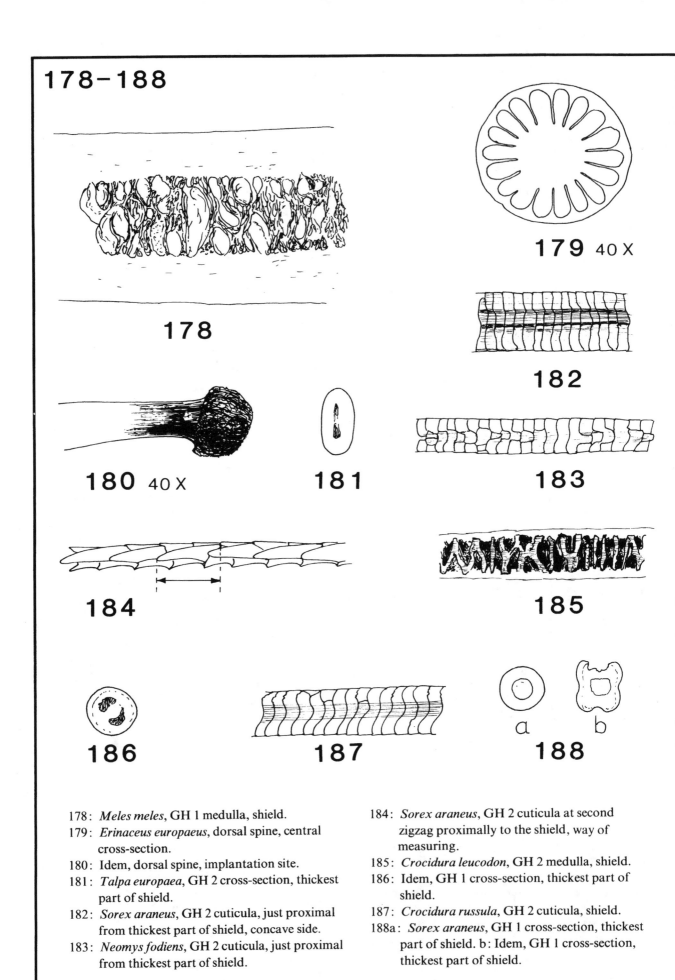

178 – 188

178

179 40 X

180 40 X

181

182

183

184

185

186

187

188

178: *Meles meles*, GH 1 medulla, shield.
179: *Erinaceus europaeus*, dorsal spine, central cross-section.
180: Idem, dorsal spine, implantation site.
181: *Talpa europaea*, GH 2 cross-section, thickest part of shield.
182: *Sorex araneus*, GH 2 cuticula, just proximal from thickest part of shield, concave side.
183: *Neomys fodiens*, GH 2 cuticula, just proximal from thickest part of shield.
184: *Sorex araneus*, GH 2 cuticula at second zigzag proximally to the shield, way of measuring.
185: *Crocidura leucodon*, GH 2 medulla, shield.
186: Idem, GH 1 cross-section, thickest part of shield.
187: *Crocidura russula*, GH 2 cuticula, shield.
188a: *Sorex araneus*, GH 1 cross-section, thickest part of shield. b: Idem, GH 1 cross-section, thickest part of shield.

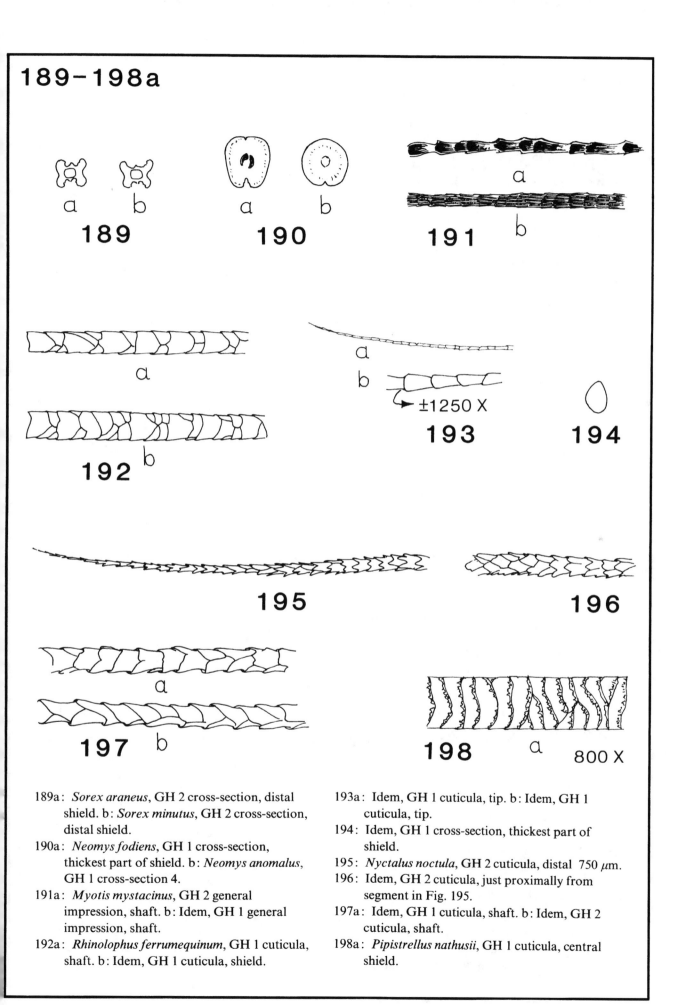

189-198a

189a: *Sorex araneus*, GH 2 cross-section, distal shield. b: *Sorex minutus*, GH 2 cross-section, distal shield.

190a: *Neomys fodiens*, GH 1 cross-section, thickest part of shield. b: *Neomys anomalus*, GH 1 cross-section 4.

191a: *Myotis mystacinus*, GH 2 general impression, shaft. b: Idem, GH 1 general impression, shaft.

192a: *Rhinolophus ferrumequinum*, GH 1 cuticula, shaft. b: Idem, GH 1 cuticula, shield.

193a: Idem, GH 1 cuticula, tip. b: Idem, GH 1 cuticula, tip.

194: Idem, GH 1 cross-section, thickest part of shield.

195: *Nyctalus noctula*, GH 2 cuticula, distal 750 μm.

196: Idem, GH 2 cuticula, just proximally from segment in Fig. 195.

197a: Idem, GH 1 cuticula, shaft. b: Idem, GH 2 cuticula, shaft.

198a: *Pipistrellus nathusii*, GH 1 cuticula, central shield.

198b–204

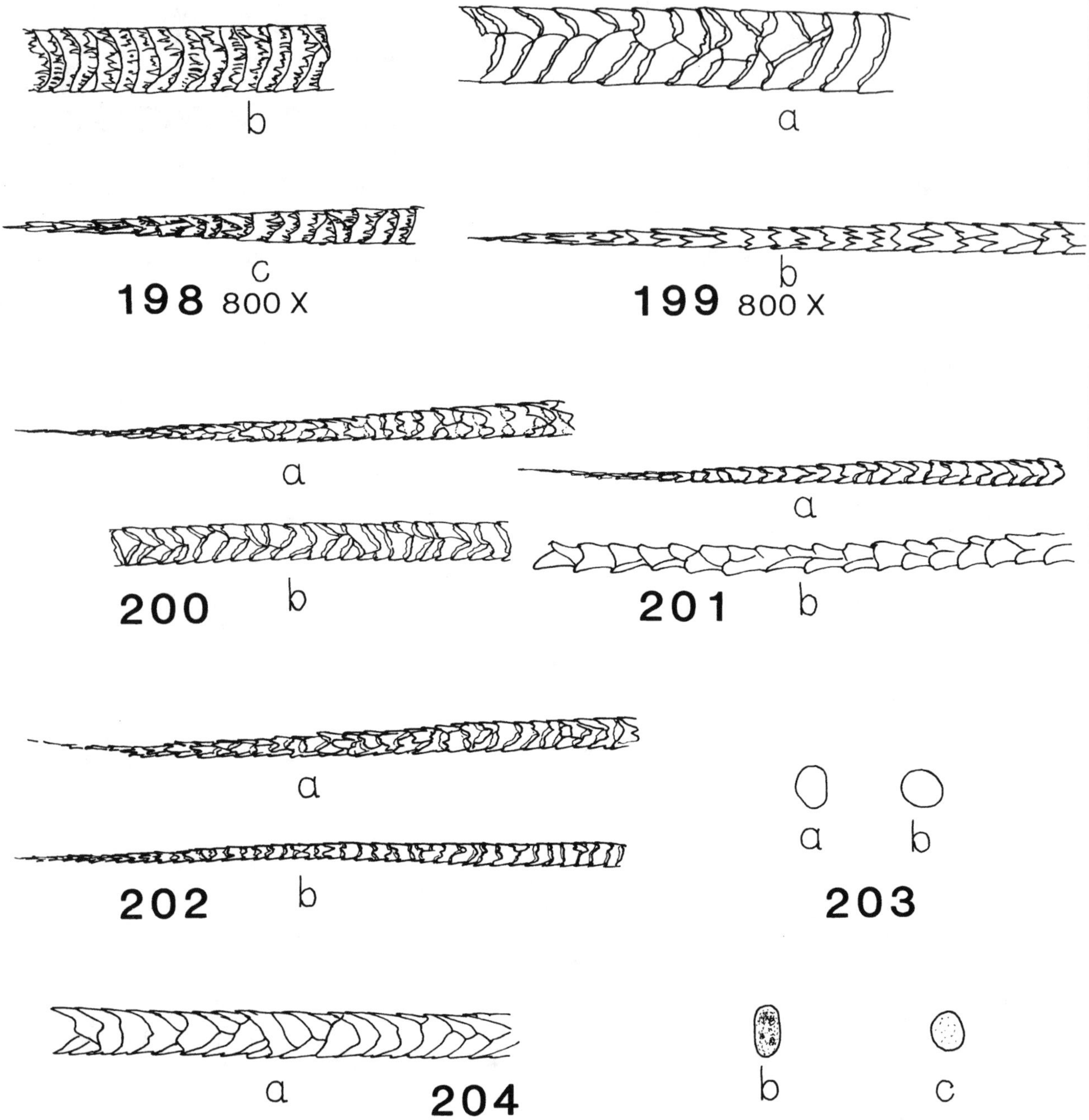

198 800 X

199 800 X

200

201

202

203

204

198b: *Pipistrellus nathusii*, GH 1 cuticula, distal shield. c: Idem, GH 1 cuticula, tip.

199a: *Pipistrellus pipistrellus*, GH 1 and GH 2 cuticula, shield. b: Idem, GH 1 and GH 2 cuticula, tip.

200a: *Nyctalus leisleri*, GH 1 cuticula, tip. b: Idem, GH 1 cuticula, thickest part of shield.

201a: *Vespertilio murinus*, GH 2 cuticula, tip. b: Idem, GH 2 cuticula, shaft.

202a: *Nyctalus leisleri*, GH 2 cuticula, tip. b:

Pipistrellus pipistrellus, GH 2 cuticula, tip.

203a: *Vespertilio murinus*, GH 1 cross-section, thickest part of shield. b: *Barbastella barbastellus*, GH 1 cross-section, thickest part of shield.

204a: *Plecotus auritus*, GH 2 and GH 1 cuticula, central shield. b: Idem, GH 1 cross-section, thickest part of shield. c: *Myotus bechsteinii*, GH 1 cross-section, thickest part of shield.

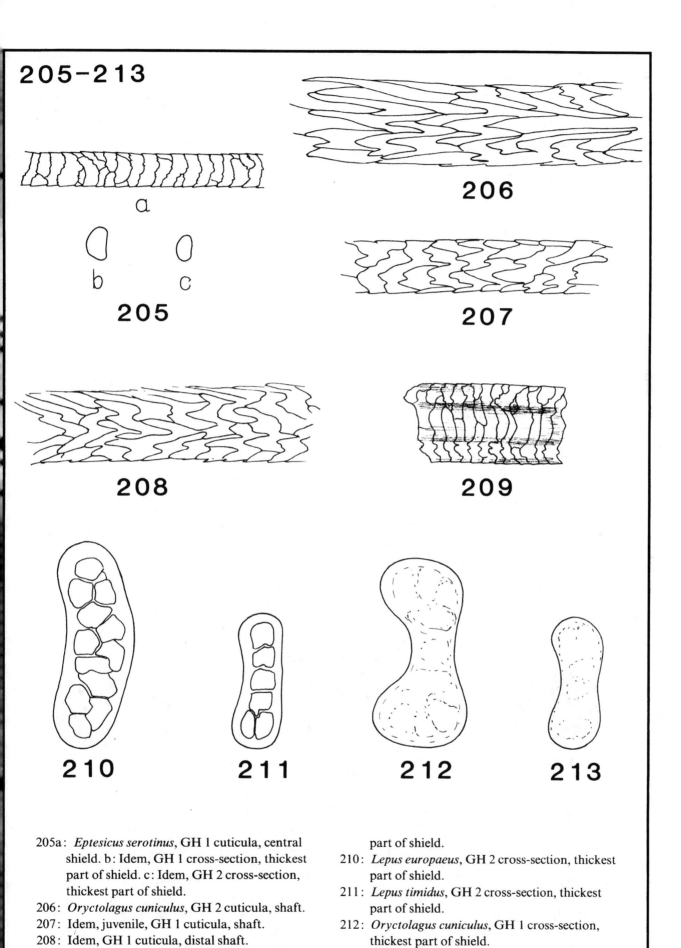

205-213

205

a

b   c

206

207

208

209

210   211   212   213

205a: *Eptesicus serotinus*, GH 1 cuticula, central
shield. b: Idem, GH 1 cross-section, thickest
part of shield. c: Idem, GH 2 cross-section,
thickest part of shield.
206: *Oryctolagus cuniculus*, GH 2 cuticula, shaft.
207: Idem, juvenile, GH 1 cuticula, shaft.
208: Idem, GH 1 cuticula, distal shaft.
209: *Apodemus agrarius*, GH 2 cuticula, thickest
part of shield.
210: *Lepus europaeus*, GH 2 cross-section, thickest
part of shield.
211: *Lepus timidus*, GH 2 cross-section, thickest
part of shield.
212: *Oryctolagus cuniculus*, GH 1 cross-section,
thickest part of shield.
213: *Lepus europaeus*, GH 2 cross-section 3.

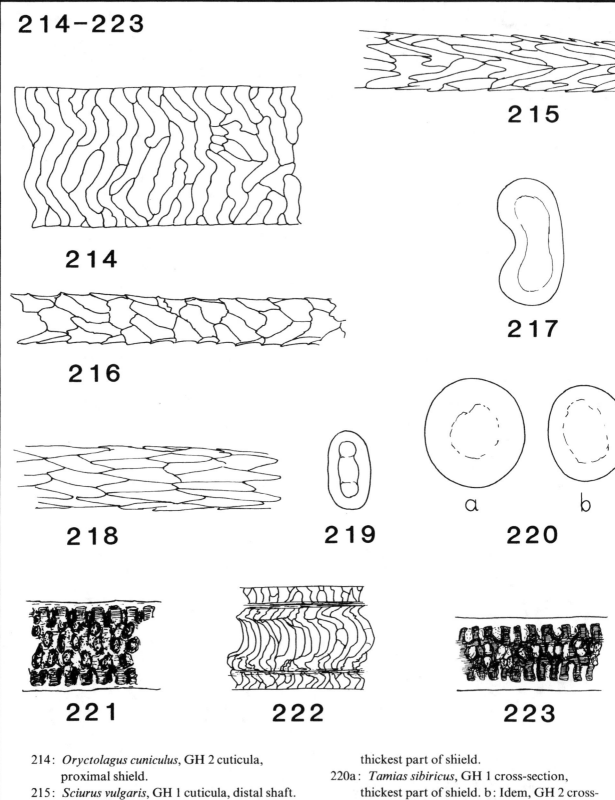

214-223

214: *Oryctolagus cuniculus*, GH 2 cuticula, proximal shield.

215: *Sciurus vulgaris*, GH 1 cuticula, distal shaft.

216: *Tamias sibiricus*, GH 1 cuticula, proximal shaft.

217: *Sciurus vulgaris*, GH 1 cross-section, thickest part of shield.

218: Idem, GH 1 cuticula, proximal and central shaft.

219: *Sciurus carolinensis*, GH 2 cross-section, thickest part of shield.

220a: *Tamias sibiricus*, GH 1 cross-section, thickest part of shield. b: Idem, GH 2 cross-section, thickest part of shield.

221: *Apodemus flavicollis*, GH 1 medulla, thickest part of shield.

222: *Clethrionomys glareolus*, GH 1 cuticula, central shield.

223: Idem, GH 2 medulla, central shield.

224-233

224   **225**   **226**

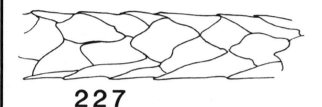

227   **228**

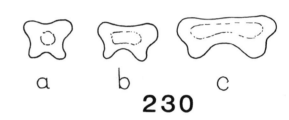

229   **230**
a   b   c

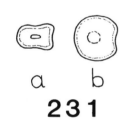

a   b
231

232

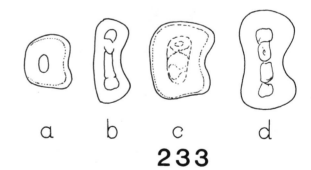

a   b   c   d
233

224: *Pitymys subterraneus*, GH 1 medulla, shaft.
225: *Arvicola terrestris*, GH 1 medulla, distal shaft.
226: *Microtus arvalis*, GH 2 medulla, distal shaft.
227: *Cricetus cricetus*, GH 1 cuticula, central shaft.
228: Idem, GH 2 cuticula, central shaft.
229: *Pitymys subterraneus*, GH 1 and GH 2, profile.
230a: *Clethrionomys glareolus*, GH2cross-section3.

b: Idem, GH 2 cross-section 4. c: Idem, GH 1 cross-section, thickest part of shield.
231a: *Pitymys subterraneus*, GH 1 cross-section 2. b: *Microtus oeconomus*, GH 1 cross-section 2.
232: Idem, GH 2 medulla, proximal shaft.
233a: *Pitymys subterraneus*, GH 1 cross-section 3. b: Idem, GH 1 cross-section 5. c: *Microtus agrestis*, GH 1 cross-section 3. d: Idem, GH 1 cross-section 5.

234-241

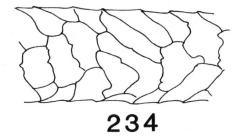

234

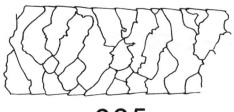

235

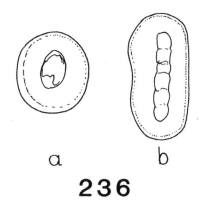

236

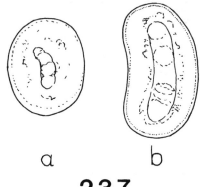

237

238

239

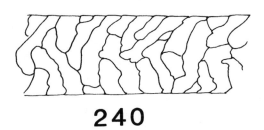

240

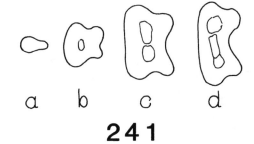

241

234: *Cricetus cricetus*, GH 1 cuticula, proximal
     shield, transitional pattern.
235: Idem, GH 1 cuticula, proximal shield just
     distally to segment in Fig. 234.
236a: *Arvicola terrestris*, GH 1 cross-section 3. b:
     Idem, GH 1 cross-section 5.
237a: *Cricetus cricetus*, GH 1 cross-section 3. b:
     Idem, GH 1 cross-section 5.

238: *Arvicola terrestris*, GH 1 cuticula, central
     shaft.
239: Idem, GH 1 cuticula, proximal shield,
     transitional pattern.
240: Idem, GH 1 cuticula, proximal shield just
     distally to segment in Fig. 239.
241a–d: *Apodemus sylvaticus*, GH 1 cross-sections
     2, 3, 5, and 6.

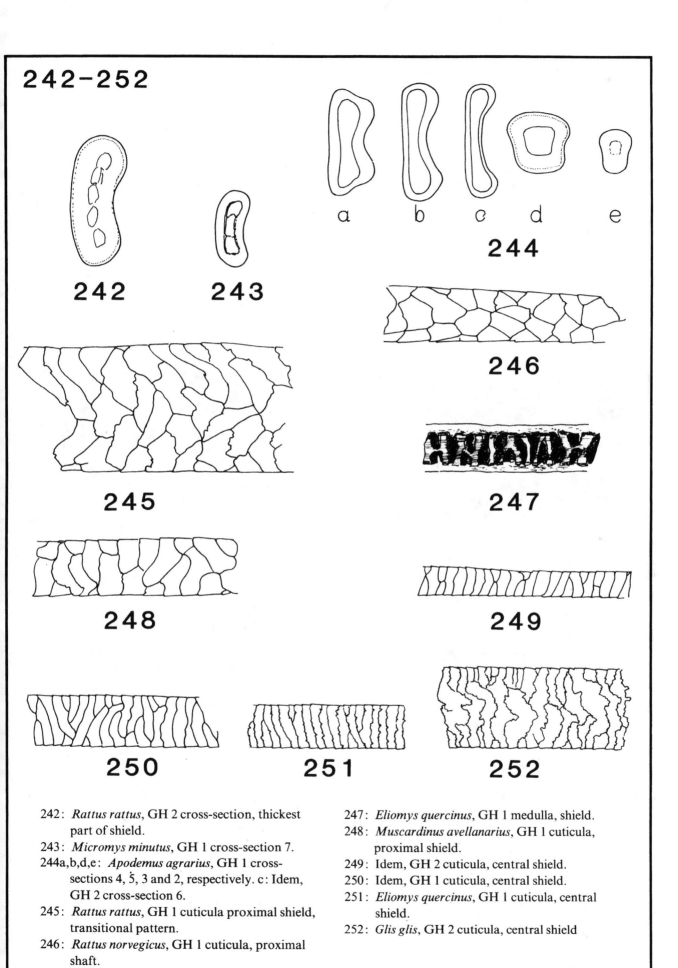

242-252

a   b   c   d   e

244

242   **243**

246

245

247

248   **249**

250   **251**   **252**

242: *Rattus rattus*, GH 2 cross-section, thickest part of shield.

243: *Micromys minutus*, GH 1 cross-section 7.

244a,b,d,e: *Apodemus agrarius*, GH 1 cross-sections 4, 5, 3 and 2, respectively. c: Idem, GH 2 cross-section 6.

245: *Rattus rattus*, GH 1 cuticula proximal shield, transitional pattern.

246: *Rattus norvegicus*, GH 1 cuticula, proximal shaft.

247: *Eliomys quercinus*, GH 1 medulla, shield.

248: *Muscardinus avellanarius*, GH 1 cuticula, proximal shield.

249: Idem, GH 2 cuticula, central shield.

250: Idem, GH 1 cuticula, central shield.

251: *Eliomys quercinus*, GH 1 cuticula, central shield.

252: *Glis glis*, GH 2 cuticula, central shield

253-259

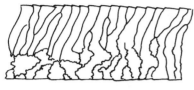

253

254

255

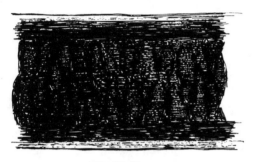

256

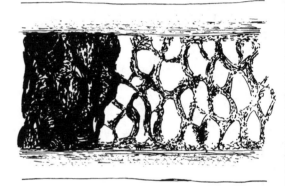

257

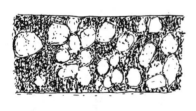

258

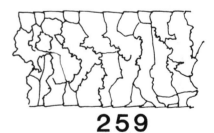

259

253: *Glis glis*, GH 1 cuticula, thickest part of shield.

254: *Eliomys quercinus*, GH 2 cross-section, thickest part of shield.

255: *Nyctereutes procyonoides*, GH 2 medulla, shaft.

256: *Canis familiaris*, GH 1 medulla, shield.

257: *Nyctereutes procyonoides*, GH 1 medulla, shield.

258: Idem, GH 1 medulla, proximal shield.

259: Idem, cuticula, central part of the hair.

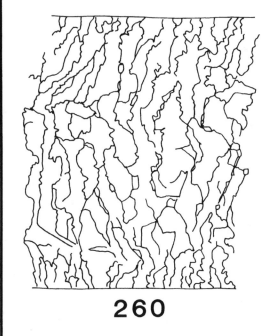

260

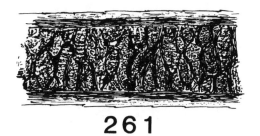

261

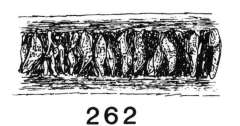

262

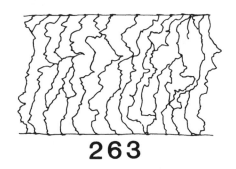

263

264

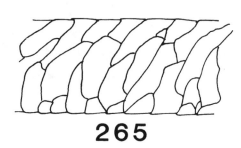

265

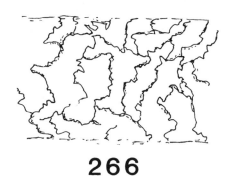

266

260: *Nyctereutes procyonoides*, GH 1 cuticula, shield.

261: *Vulpes vulpes*, GH 2 medulla, shaft.

262: Idem, GH 2 medulla, shaft.

263: Idem, GH 2 cuticula, shield.

264: *Canis familiaris*, GH 2 medulla, proximal shield.

265: Idem, GH 2 cuticula, curved part of the shaft.

266: *Meles meles*, GH 1 cuticula, distal shaft.

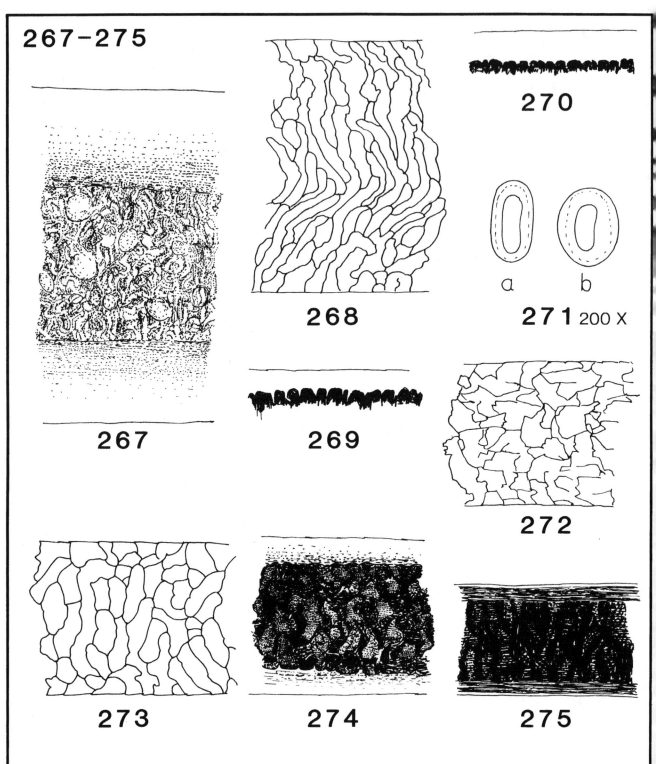

267–275

267

268

269

270

271 200 X

a   b

272

273

274

275

267: *Meles meles*, GH 1 medulla, distal half of the hair.
268: Idem, GH 1 cuticula, near proximal shield.
269: *Mustela nivalis*, GH 1 and GH 2 medulla border, thickest part of shield, no oil penetration.
270: *Mustela erminea*, GH 1 and GH 2 medulla, thickest part of shield, no oil penetration.
271a: *Mustela nivalis*, GH 1 cross-section, thickest part of shield. b: *Mustela erminea*, GH 1 cross-section, thickest part of shield.
272: *Lutra lutra*, GH 1 cuticula, distal shaft, proximal shield.
273: *Mustela vison*, GH 1 cuticula, proximal shield.
274: *Martes foina*, GH 1 medulla just distally to middle part of shield.
275: *Mustela putorius*, GH 1 medulla, distal shield.

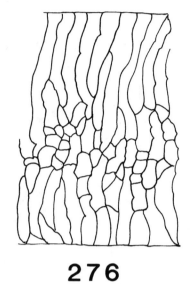

276

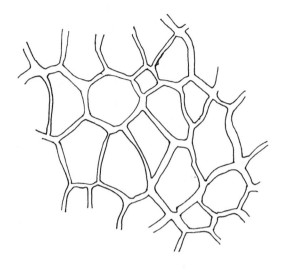

277

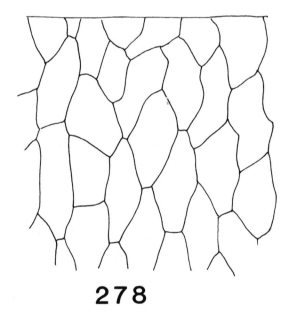

278

279 nat.size

280

276: *Lutra lutra*, GH 2 cuticula, shield.
277: *Cervus elaphus*, GH 1 (long hair), medulla, at $\frac{1}{4}$ distance from base (oil penetration).
278: *Capreolus capreolus*, GH 1 cuticula, at $\frac{2}{3}$ from tip.

279: *Sus scrofa*, GH 1 hair profile, tip.
280: Way of counting the turns in artiodactylar hairs.

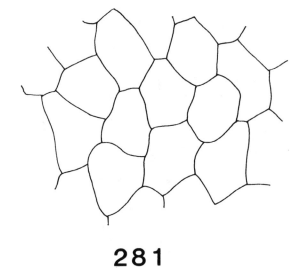

281

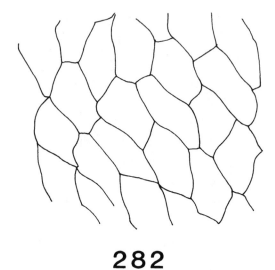

282

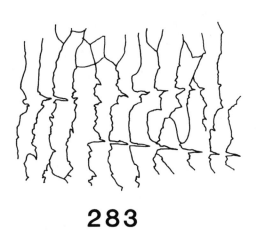

283

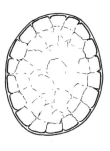

284 100 ×

281: *Ovis musimon*, GH 1 cuticula, at base.
282: *Cervus dama*, GH 1 cuticula, at base.
283: *Capreolus capreolus*, GH 1 cuticula, distal part.

284: Idem, GH 1 cross-section, thickest part of shield.

Part III
Atlas

7 Introduction

In this atlas section, all species are numbered and classified according to orders and families. Each species is given two pages. The various types of hair with the corresponding magnification standards are shown just below the species name on the lefthand page. With few exceptions, only one standard is used within orders. On this basis, differences in size can be seen at a glance.

Cross-sections of GH 1 and GH 2 are represented in the left and right blocks on the lefthand page, respectively. The numbers along the schematically represented hair indicate the places from which the sections (given the same numbers) were taken. In these drawings, the tip of the hair points to the left, as in all of the Figures and photographs in this publication. Approximate ratios as relative length of shaft and shield can be derived from these schematic hairs, which are all of the same length. The standard magnification of the cross-sections is recorded at the top of the block to indicate the image under the microscope. Exact figures can be derived from a bar divided into microns.

On the right page, photographs of cuticula and medulla are shown. The positions of GH 1 and GH 2 patterns on the hairs are shown for cuticular characters in capitals above, for medullar characters underneath the drawings.

In some cases, the multiplicity of appearances in a single type of hair made it impossible to show the entire range of characteristics in photographs, but additional information can often be found in the keys.

The width of a pattern in this atlas section does not always represent the total width at that place. There are two reasons for this. First, where margins were out of focus and blurred in the photographs they were cut off, and second, hairs were sometimes too large for the chosen scale. In the latter case, however, the sizes of the cells in the pictures permits acccurate comparison with the images under the microscope (see also section 5.3).

Note
1. Because hairs of *Chiroptera* have no medulla, this heading does not appear above the relevant photographs. The term *General impression* is used instead of *Medulla*. Although the slides and photographs were made in exactly the same way as medullar ones, they are less useful for identification.
2. Cross-sections of zigzag shafts such as occur in *Insectivora*, were normally taken in the middle of a straight part.

1 Erinaceus europaeus

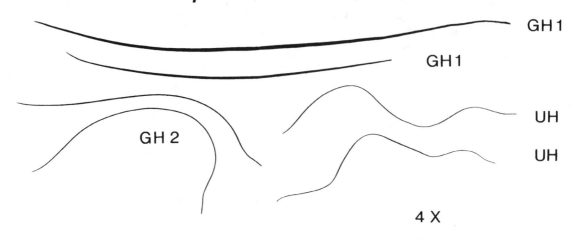

GH 1
GH 1
GH 2
UH
UH

4 X

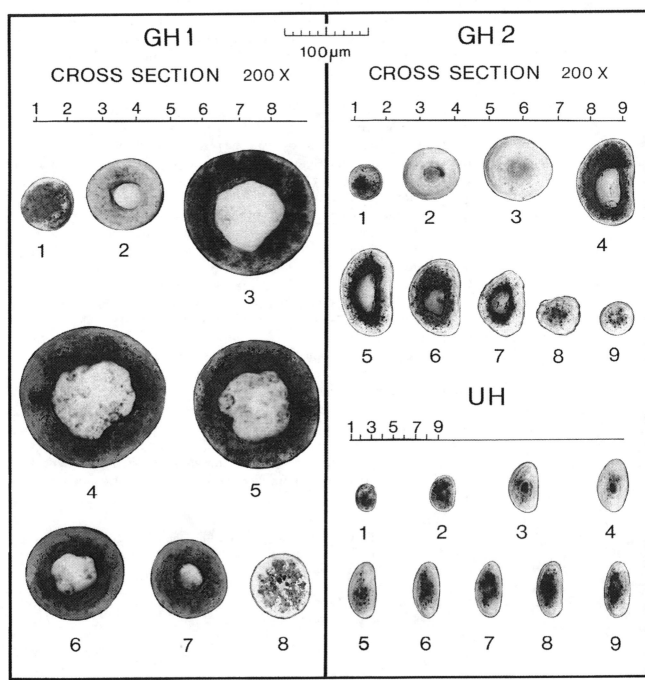

GH 1
CROSS SECTION 200 X
100 µm

GH 2
CROSS SECTION 200 X

UH

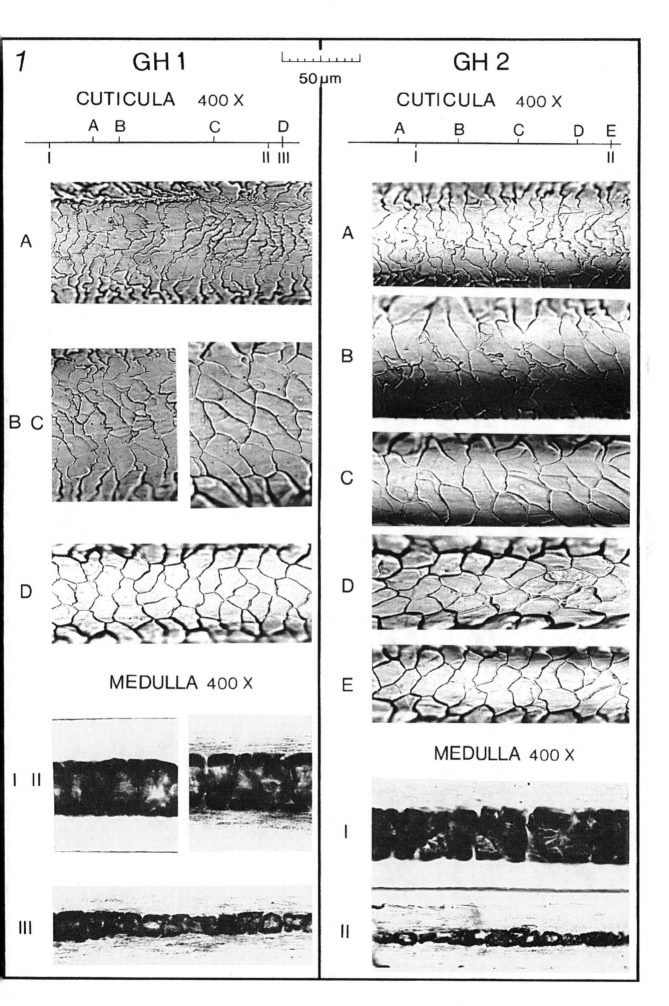

1

GH 1

CUTICULA   400 X

A   B   C   D
I   II III

A

B C

D

MEDULLA 400 X

I II

III

GH 2

CUTICULA   400 X

A   B   C   D E
I   II

A

B

C

D

E

MEDULLA 400 X

I

II

50 µm

2 Sorex araneus

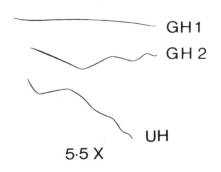

GH 1
GH 2
UH
5·5 X

GH 1	GH 2
CROSS SECTION 400 X	CROSS SECTION 400 X

50 µm

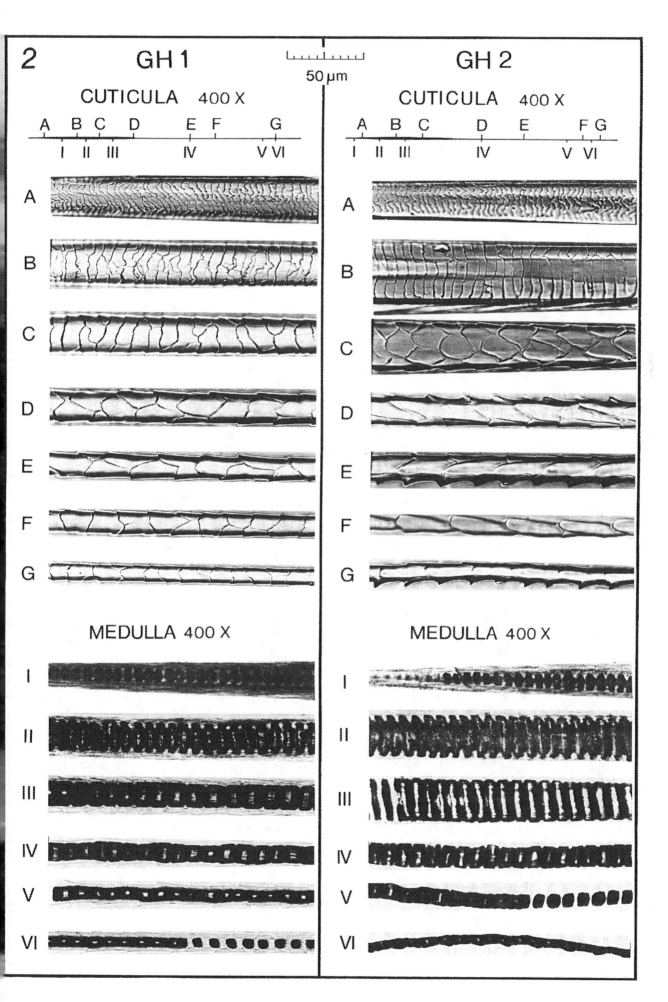

2 GH 1   GH 2

CUTICULA 400 X   CUTICULA 400 X

MEDULLA 400 X   MEDULLA 400 X

75

3 Sorex minutus

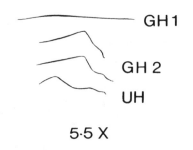

GH 1

GH 2

UH

5·5 X

GH 1 | 50 µm | GH 2

CROSS SECTION 400 X | CROSS SECTION 400 X

1 2 3 4 5 6 7 8 9   10 11 | 1 2 3 4 5 6 7 8 9   10 11

1 2 3

4 5 6

7 8 9

10   11

1 2 3

4 5 6

7 8 9

10   11

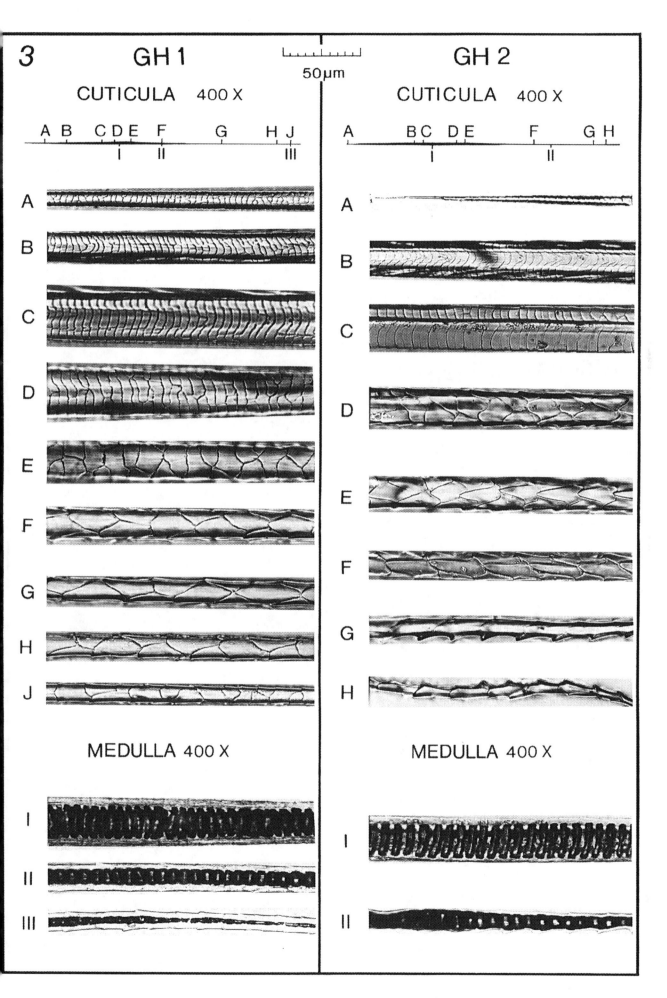

3 **GH 1** 50µm **GH 2**

CUTICULA 400 X CUTICULA 400 X

4 Neomys fodiens

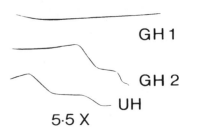

GH 1

GH 2

UH

5·5 X

GH 1

CROSS SECTION 400 X

1 3 5 7 9 10 11

50 μm

GH 2

CROSS SECTION 400 X

1 3 5 7 9 10 11

1   2   3

4   5   6

7   8   9

10   11

1   2   3

4   5   6

7   8   9

10   11

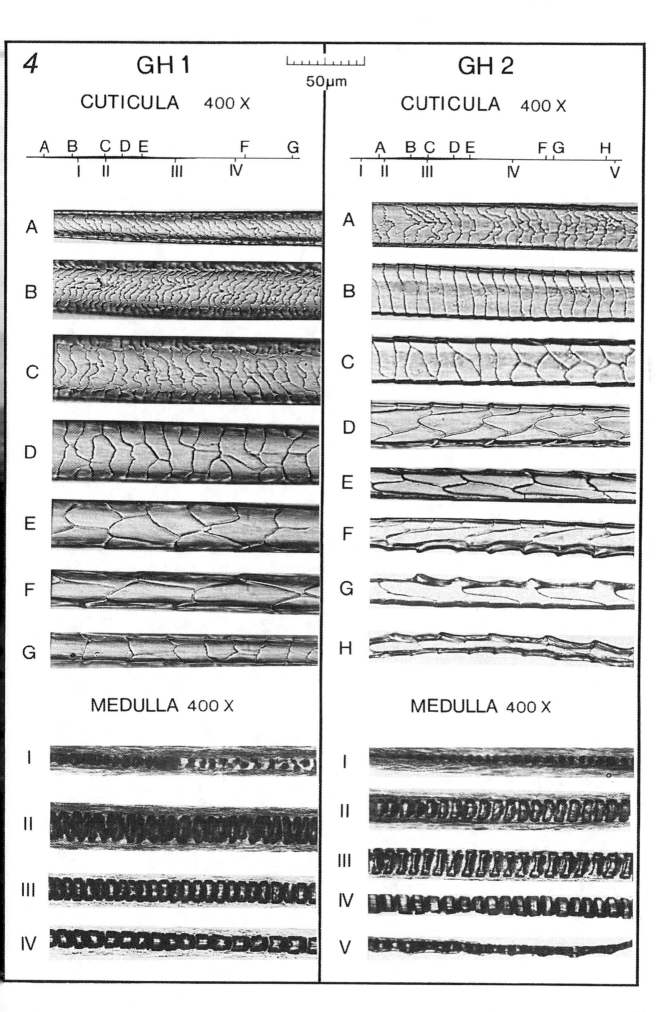

4　GH 1

CUTICULA　400 X

A　B　C D E　F　G
I　II　III　IV

A
B
C
D
E
F
G

MEDULLA 400 X

I
II
III
IV

GH 2

CUTICULA　400 X

A　B C D E　F G　H
I　II　III　IV　V

A
B
C
D
E
F
G
H

MEDULLA 400 X

I
II
III
IV
V

50µm

5 Neomys anomalus

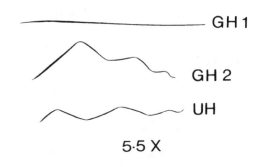

GH 1

GH 2

UH

5·5 X

GH 1	GH 2
CROSS SECTION 400 X	CROSS SECTION 400 X

50 µm

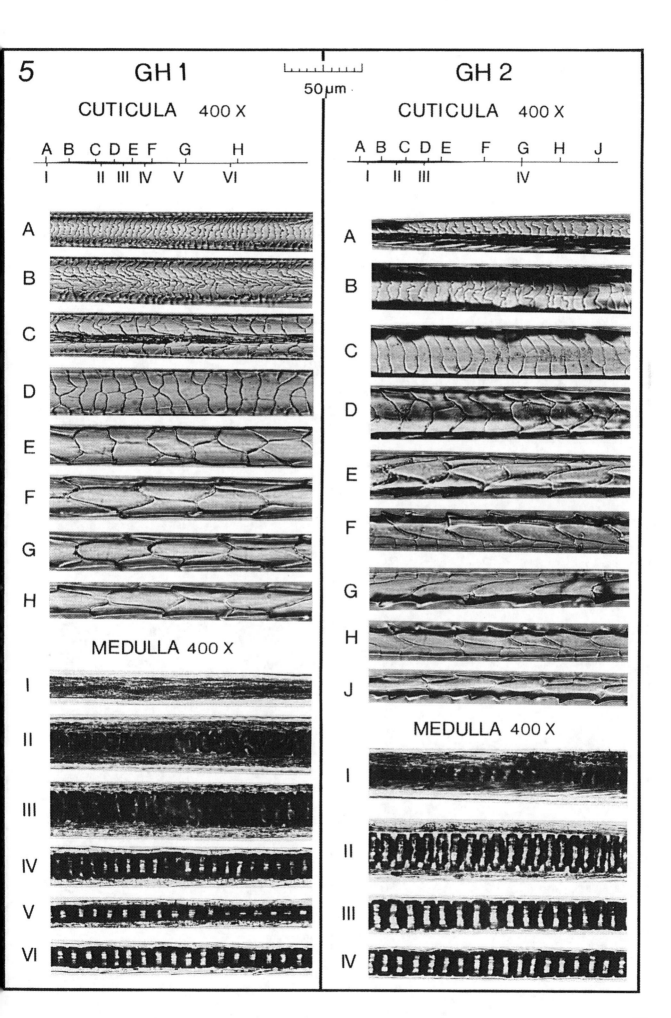

5 **GH 1**

CUTICULA 400 X

A B C D E F G H

I II III IV V VI

GH 2

CUTICULA 400 X

A B C D E F G H J

I II III IV

50µm

MEDULLA 400 X

MEDULLA 400 X

81

6 Crocidura russula

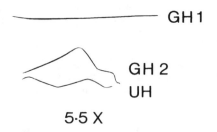

GH 1

GH 2
UH

5·5 X

GH 1   50 µm   GH 2

CROSS SECTION 400 X   CROSS SECTION 400 X

1 3 5 7 9 10 11   1 3 5 7 9 10 11

1 2 3   1 2 3

4 5 6   4 5 6

7 8 9   7 8 9

10 11   10 11

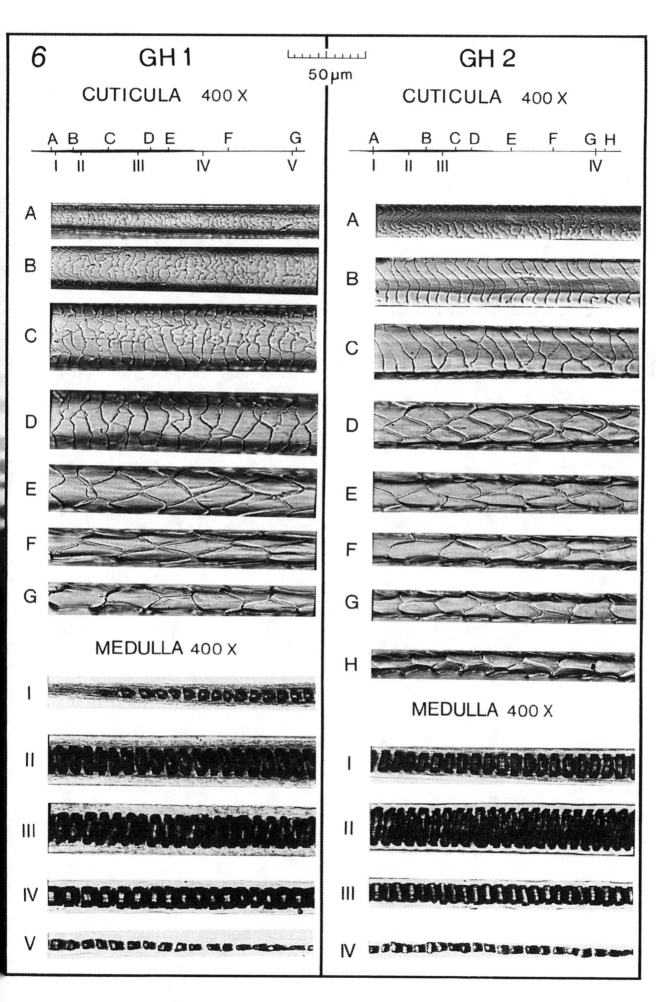

6   **GH 1**

CUTICULA 400 X

A B C D E F G

I II III IV V

A

B

C

D

E

F

G

MEDULLA 400 X

I

II

III

IV

V

GH 2

CUTICULA 400 X

A B C D E F G H

I II III IV

A

B

C

D

E

F

G

H

MEDULLA 400 X

I

II

III

IV

7 Crocidura suaveolens

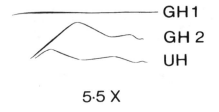

GH1
GH2
UH

5·5 X

GH 1		GH 2

GH 1

CROSS SECTION 400 X

50 µm

GH 2

CROSS SECTION 400 X

1 3 5 7 9 10 11

1 3 5 7 9 10 11

1 2 3

4 5 6

7 8 9

10 11

1 2 3

4 5 6

7 8 9

10 11

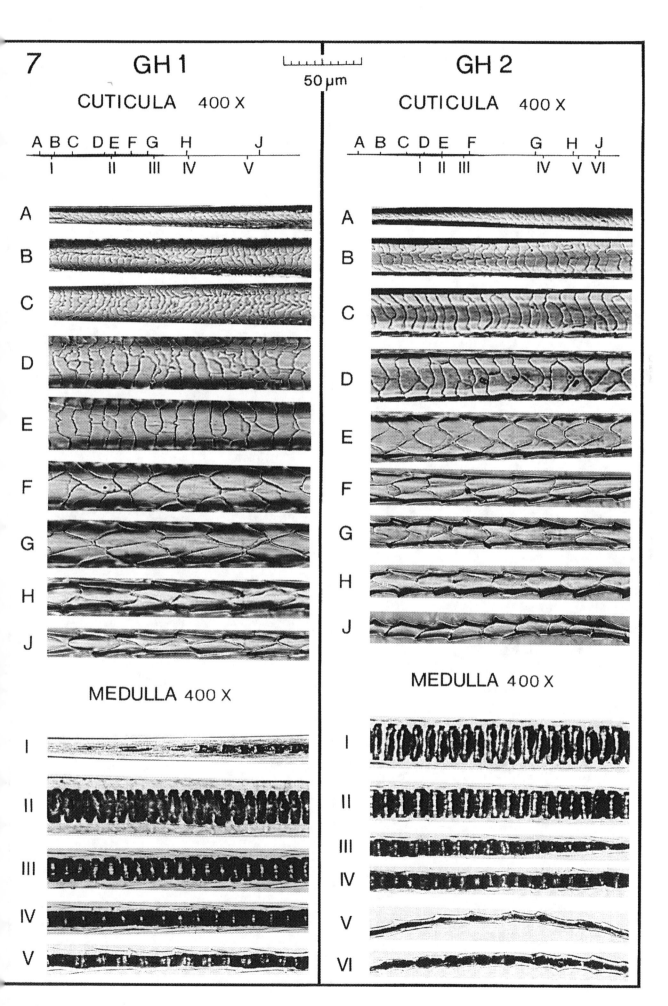

85

8 *Crocidura leucodon*

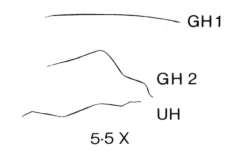

GH 1

GH 2

UH

5·5 X

GH 1

CROSS SECTION 400 X

50 μm

1 3 5 7 9 10 11

1

2

3

4

5

6

7

8

9

10

11

GH 2

CROSS SECTION 400 X

1 3 5 7 9 10 11

1

2

3

4

5

6

7

8

9

10

11

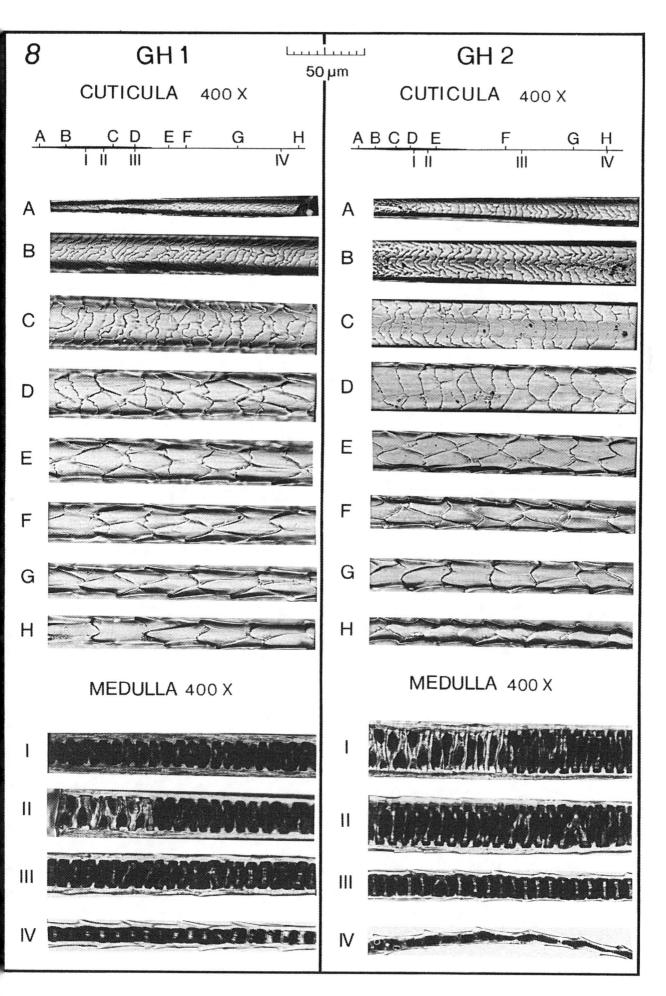

8   GH 1   50 µm   GH 2

CUTICULA 400 X   CUTICULA 400 X

A B   C D E F   G   H   A B C D E   F   G   H
      I II III   IV   I II   III   IV

A   A

B   B

C   C

D   D

E   E

F   F

G   G

H   H

MEDULLA 400 X   MEDULLA 400 X

I   I

II   II

III   III

IV   IV

9 Talpa europaea

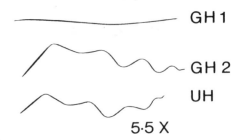

GH 1
GH 2
UH
5·5 X

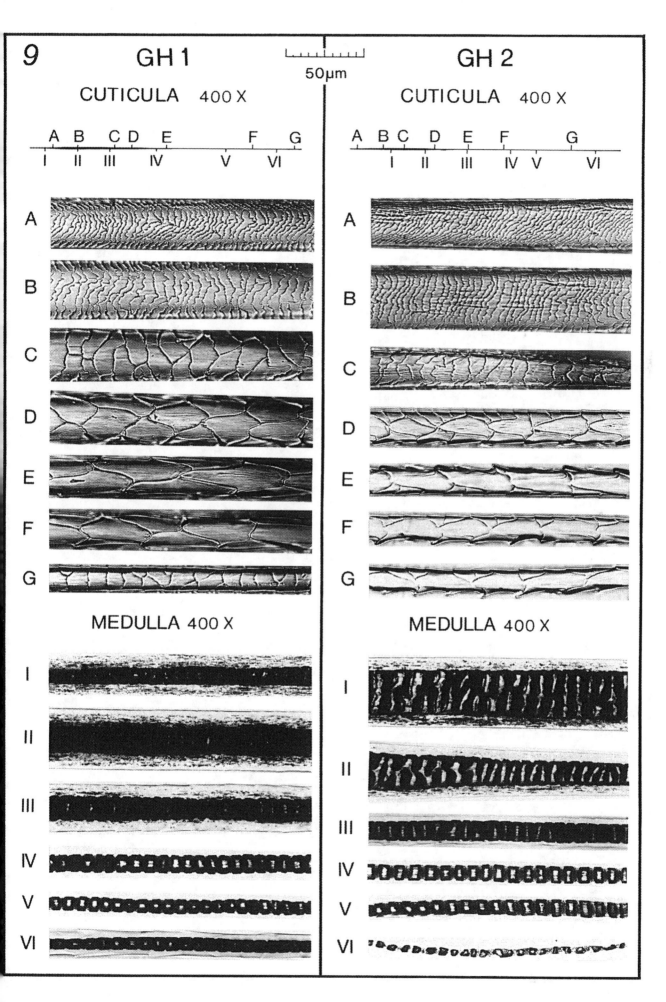

9

GH 1

CUTICULA 400 X

50µm

A B C D E F G
I II III IV V VI

A
B
C
D
E
F
G

MEDULLA 400 X

I
II
III
IV
V
VI

GH 2

CUTICULA 400 X

A B C D E F G
I II III IV V VI

A
B
C
D
E
F
G

MEDULLA 400 X

I
II
III
IV
V
VI

10 Rhinolophus ferrumequinum

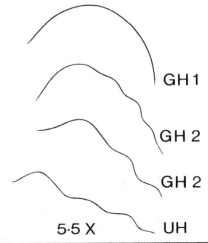

GH 1
GH 2
GH 2
5·5 X
UH

GH 1

CROSS SECTION 400 X

| 1 | 3 | 5 | 7 | 9 | 10 | 11 |

50 µm

GH 2

CROSS SECTION 400 X

| 1 | 3 | 5 | 7 | 9 | 10 | 11 |

GH 1 specimens: 1 2 3 4 5 6 7 8 9 10 11

GH 2 specimens: 1 2 3 4 5 6 7 8 9 10 11

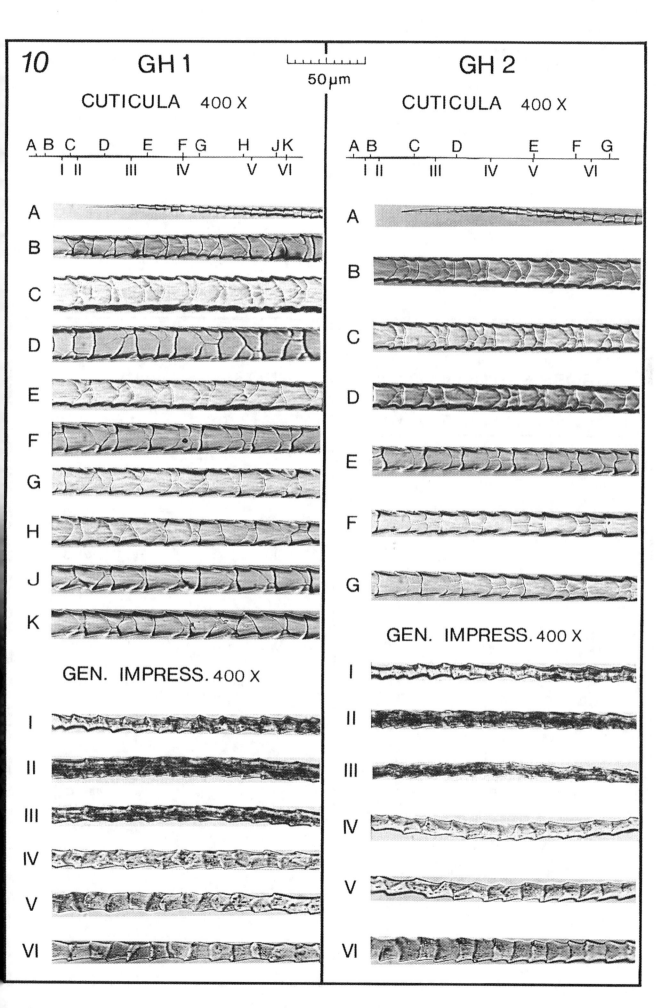

11 Rhinolophus hipposideros

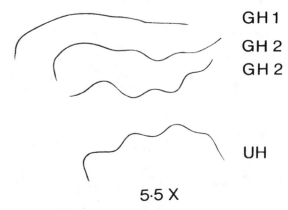

GH 1
GH 2
GH 2

UH

5·5 X

GH 1	GH 2
CROSS SECTION 400 X	CROSS SECTION 400 X

50 μm

GH 1

1   2   3

4   5   6

7   8   9

10   11

GH 2

1   2   3

4   5   6

7   8   9

10   11

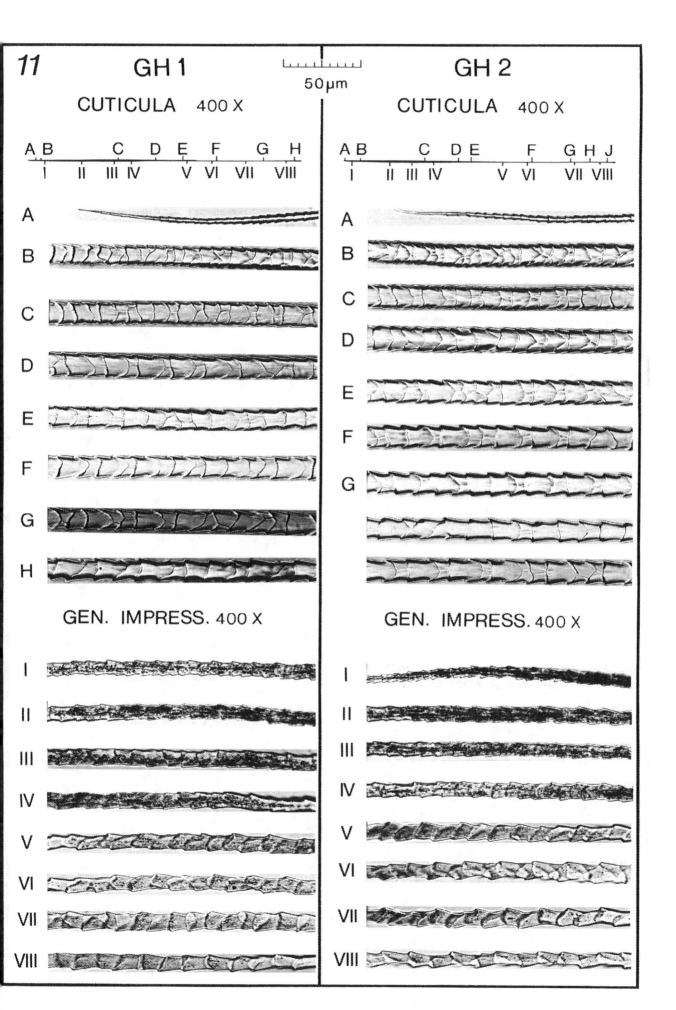

11

GH 1

CUTICULA 400 X

GH 2

CUTICULA 400 X

50µm

12 *Myotis mystacinus*

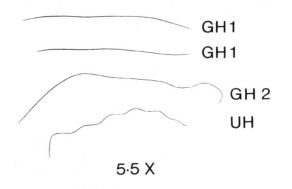

GH1

GH1

GH2

UH

5·5 X

GH1		GH2	
CROSS SECTION 400 X		CROSS SECTION 400 X	

50µm

GH1 scale: 1 3 5 7 9 10 11

GH2 scale: 1 3 5 7 9 10 11

1 2 3

4 5 6

7 8 9

10 11

1 2 3

4 5 6

7 8 9

10 11

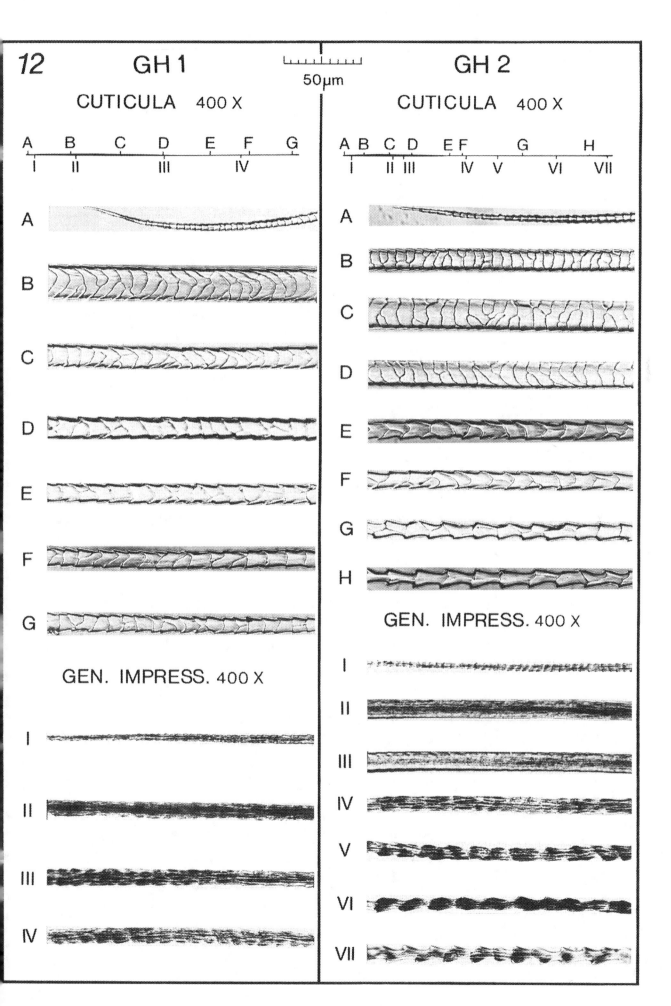

95

13 Myotis brandtii

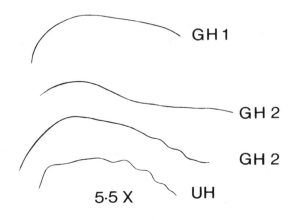

GH 1

GH 2

GH 2

5·5 X   UH

GH 1

CROSS SECTION   400 X

50 µm

| 1 | 3 | 5 | 7 | 9 | | 10 | 11 |

1   2   3

4   5   6

7   8   9

10   11

GH 2

CROSS SECTION   400 X

| 1 | 3 | 5 | 7 | 9 | | 10 | 11 |

1   2   3

4   5   6

7   8   9

10   11

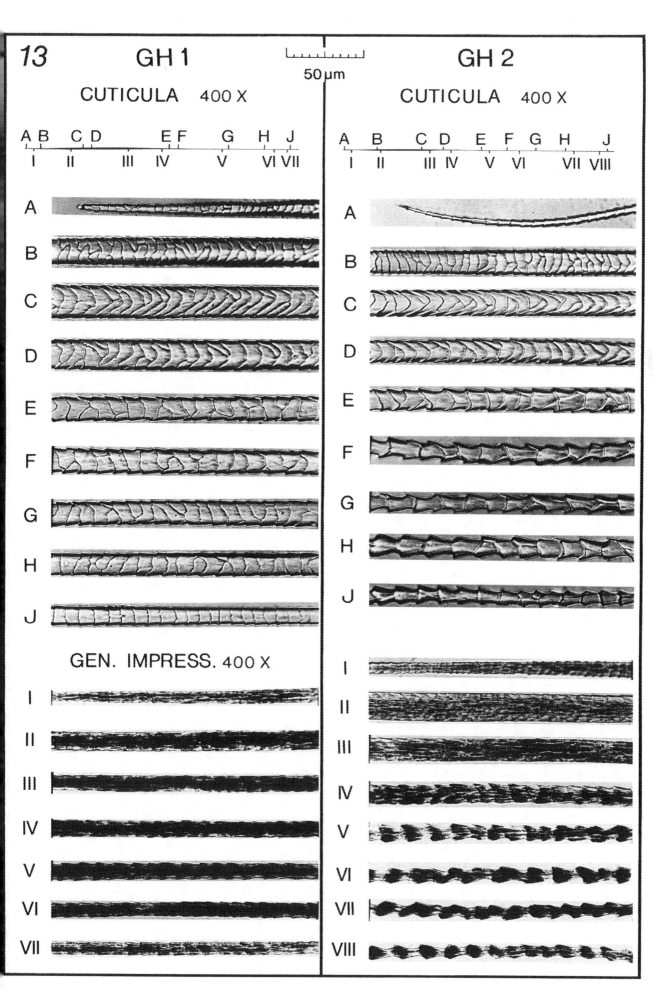

13 GH 1
CUTICULA 400 X

50 µm

GH 2
CUTICULA 400 X

GEN. IMPRESS. 400 X

14 Myotis emarginatus

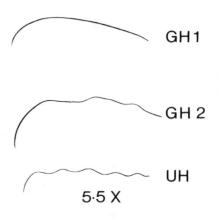

GH 1

GH 2

UH

5·5 X

GH 1	GH 2
CROSS SECTION 400 X	CROSS SECTION 400 X

50 μm

GH 1

CROSS SECTION 400 X

1 3 5 7 9   10 11

1   2   3

4   5   6

7   8   9

10   11

GH 2

CROSS SECTION 400 X

1 3 5 7 9   10 11

1   2   3

4   5   6

7   8   9

10   11

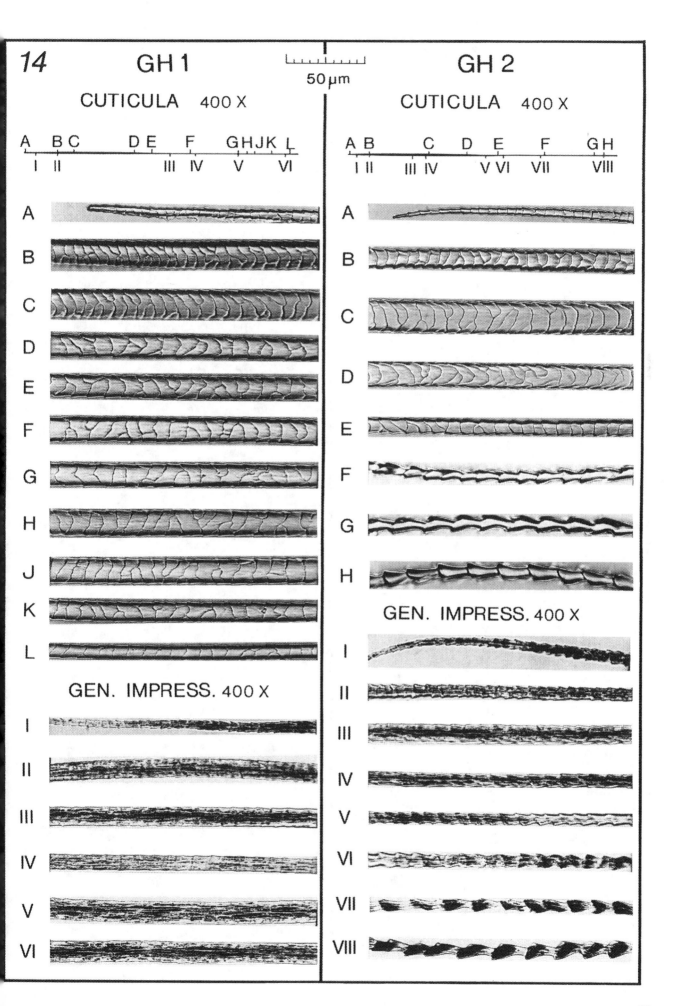

15 Myotis nattereri

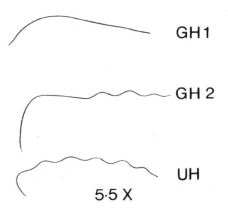

GH1

GH 2

UH

5·5 X

GH 1 | 50µm | GH 2

CROSS SECTION 400 X

CROSS SECTION 400 X

1 3 5 7 9   10 11

1 3 5 7 9   10   11

1   2   3

1   2   3

4   5   6

4   5   6

7   8   9

7   8   9

10   11

10   11

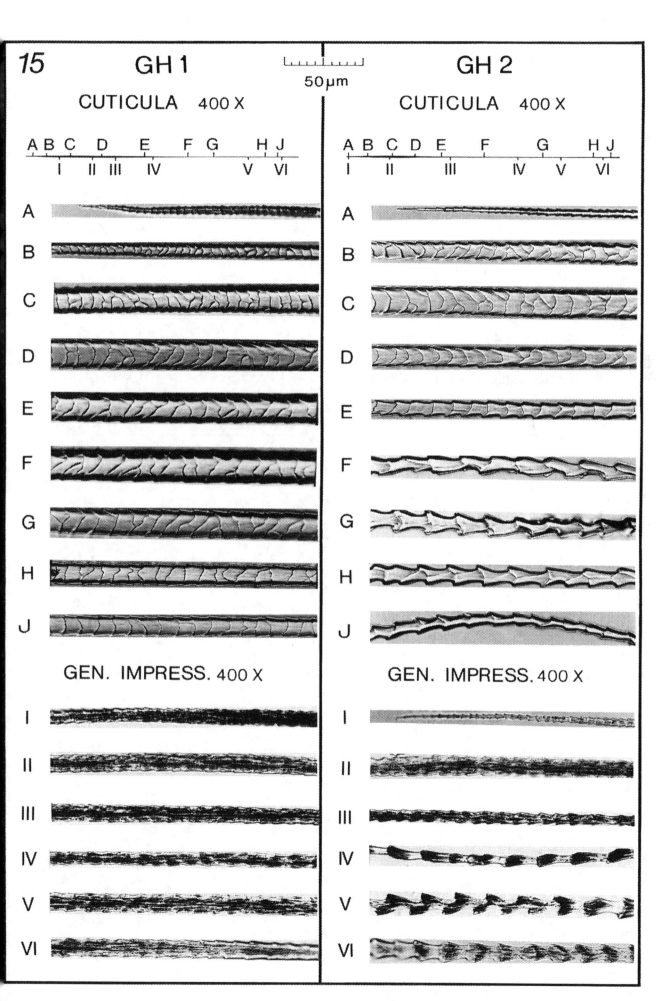

15 GH 1 CUTICULA 400 X

50μm

GH 2 CUTICULA 400 X

GEN. IMPRESS. 400 X

GEN. IMPRESS. 400 X

16 Myotis bechsteinii

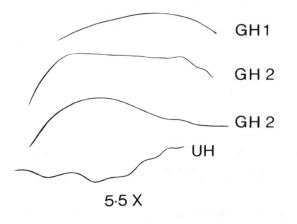

GH 1

GH 2

GH 2

UH

5·5 X

GH 1	50 µm	GH 2
CROSS SECTION 400 X		CROSS SECTION 400 X

GH 1

CROSS SECTION   400 X

1 3 5 7 9   10   11

1   2   3

4   5   6

7   8   9

10   11

GH 2

CROSS SECTION   400 X

1 3 5 7 9   10   11

1   2   3

4   5   6

7   8   9

10   11

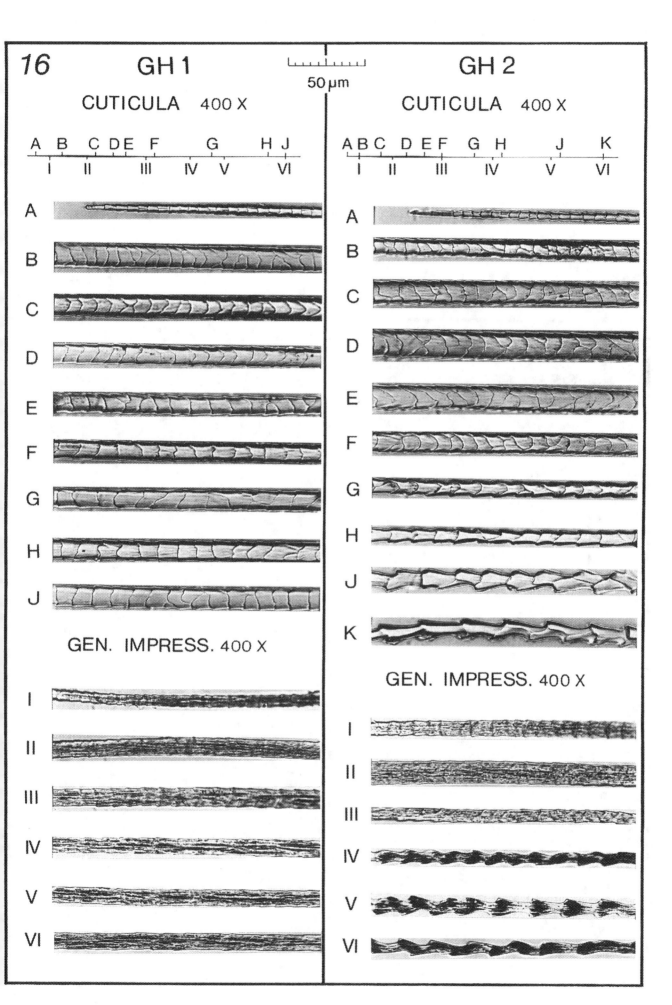

16 GH 1   50 µm   GH 2

CUTICULA 400 X   CUTICULA 400 X

GEN. IMPRESS. 400 X

GEN. IMPRESS. 400 X

103

17 Myotis myotis

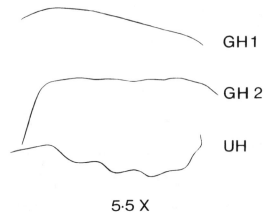

GH 1

GH 2

UH

5·5 X

GH 1

50 µm

CROSS SECTION 400 X

1 3 5 7 9   10 11

1 2 3

4 5 6

7 8 9

10   11

GH 2

CROSS SECTION 400 X

1 3 5 7 9   10 11

1 2 3

4 5 6

7 8 9

10   11

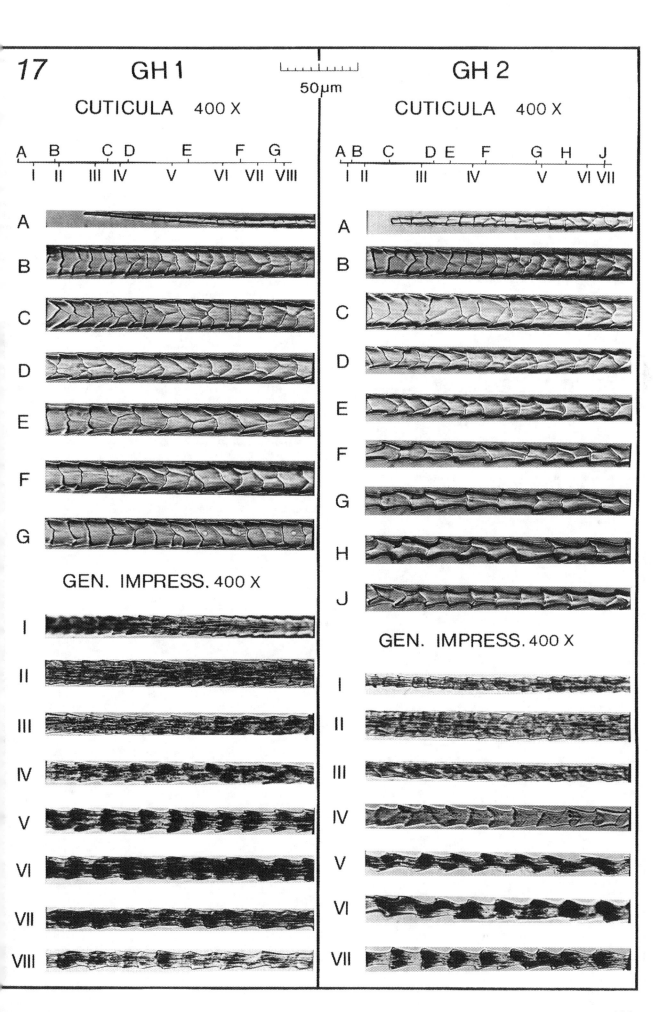

18 Myotis daubentonii

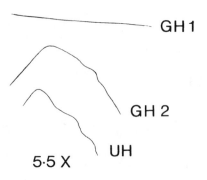

GH1

GH2

UH

5·5 X

GH1	50 μm	GH2
CROSS SECTION 400 X		CROSS SECTION 400 X

GH1 CROSS SECTION 400 X

1 3 5 7 9 10 11

1   2   3

4   5   6

7   8   9

10   11

GH2 CROSS SECTION 400 X

1 3 5 7 9 10 11

1   2   3

4   5   6

7   8   9

10   11

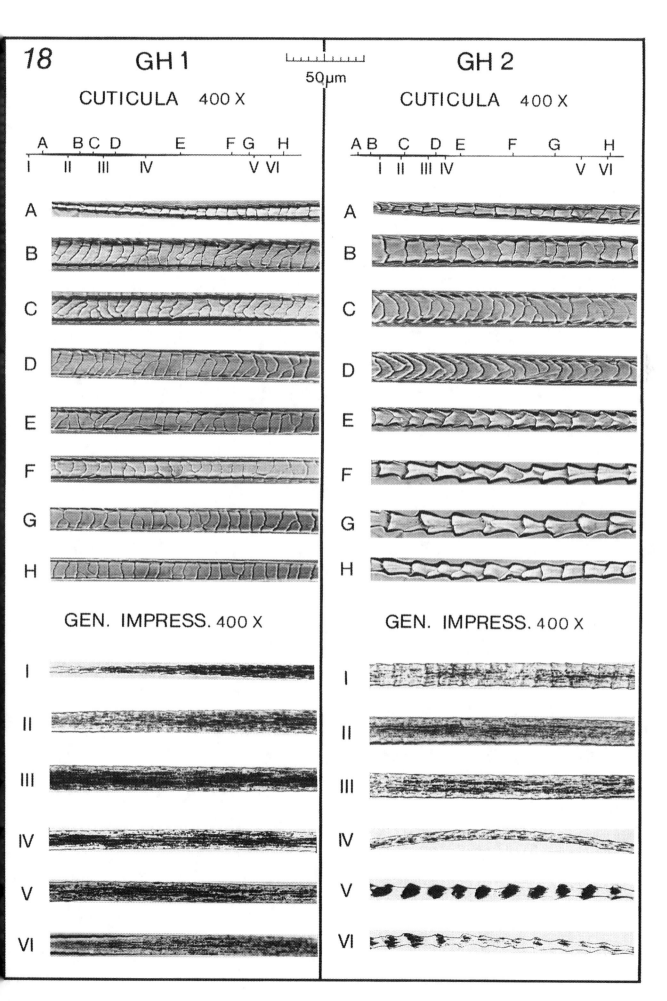

18 GH 1

CUTICULA 400 X

50µm

GH 2

CUTICULA 400 X

GEN. IMPRESS. 400 X

GEN. IMPRESS. 400 X

19 Myotis dasycneme

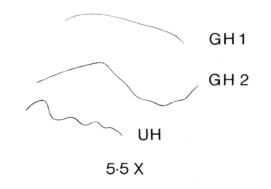

GH 1

GH 2

UH

5·5 X

GH 1		50 µm	GH 2	

GH 1

CROSS SECTION 400 X

1 3 5 7 9   10 11

1   2   3

4   5   6

7   8   9

10   11

GH 2

CROSS SECTION 400 X

1 3 5 7 9   10 11

1   2   3

4   5   6

7   8   9

10   11

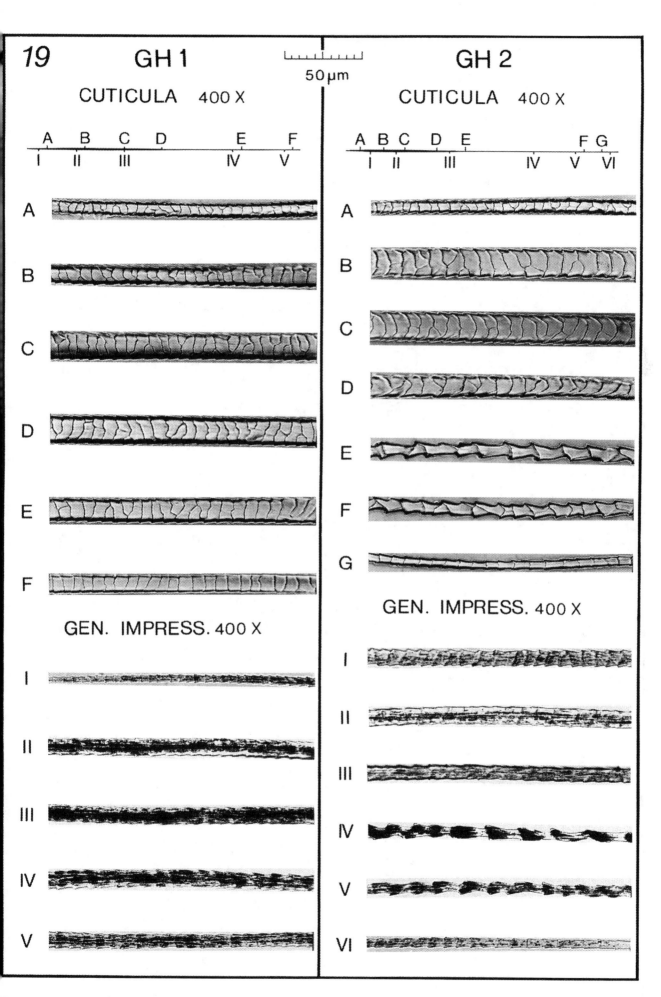

20 Pipistrellus pipistrellus

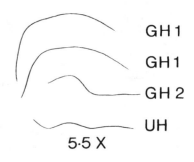

GH 1
GH 1
GH 2
UH

5·5 X

GH 1		GH 2

50 µm

CROSS SECTION 400 X   CROSS SECTION 400 X

1 3 5 7 9   10   11   1 3 5 7 9   10   11

1   2   3   1   2   3

4   5   6   4   5   6

7   8   9   7   8   9

10   11   10   11

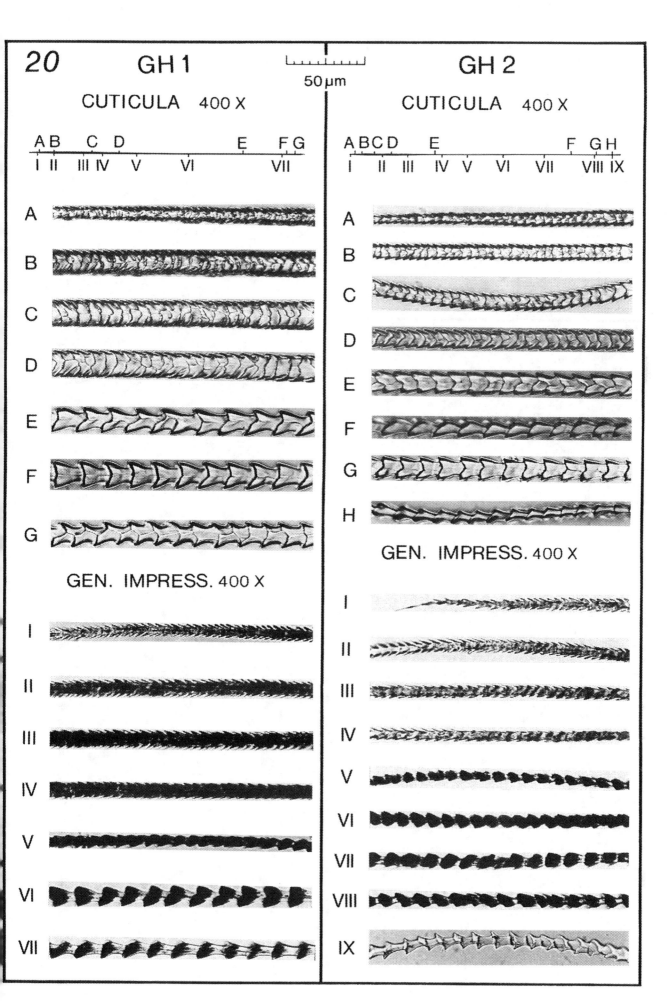

20   GH 1   50 µm   GH 2

CUTICULA 400 X   CUTICULA 400 X

A B   C   D   E   F G   A BCD   E   F G H
I II   III IV   V   VI   VII   I   II   III   IV V   VI   VII   VIII IX

A   A

B   B

C   C

D   D

E   E

F   F

G   G

                               H

GEN. IMPRESS. 400 X   GEN. IMPRESS. 400 X

I   I

II   II

III   III

IV   IV

V   V

VI   VI

VII   VII

                               VIII

                               IX

21 Pipistrellus nathusii

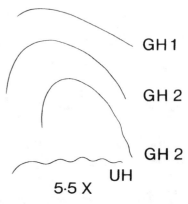

GH 1

GH 2

GH 2

UH

5·5 X

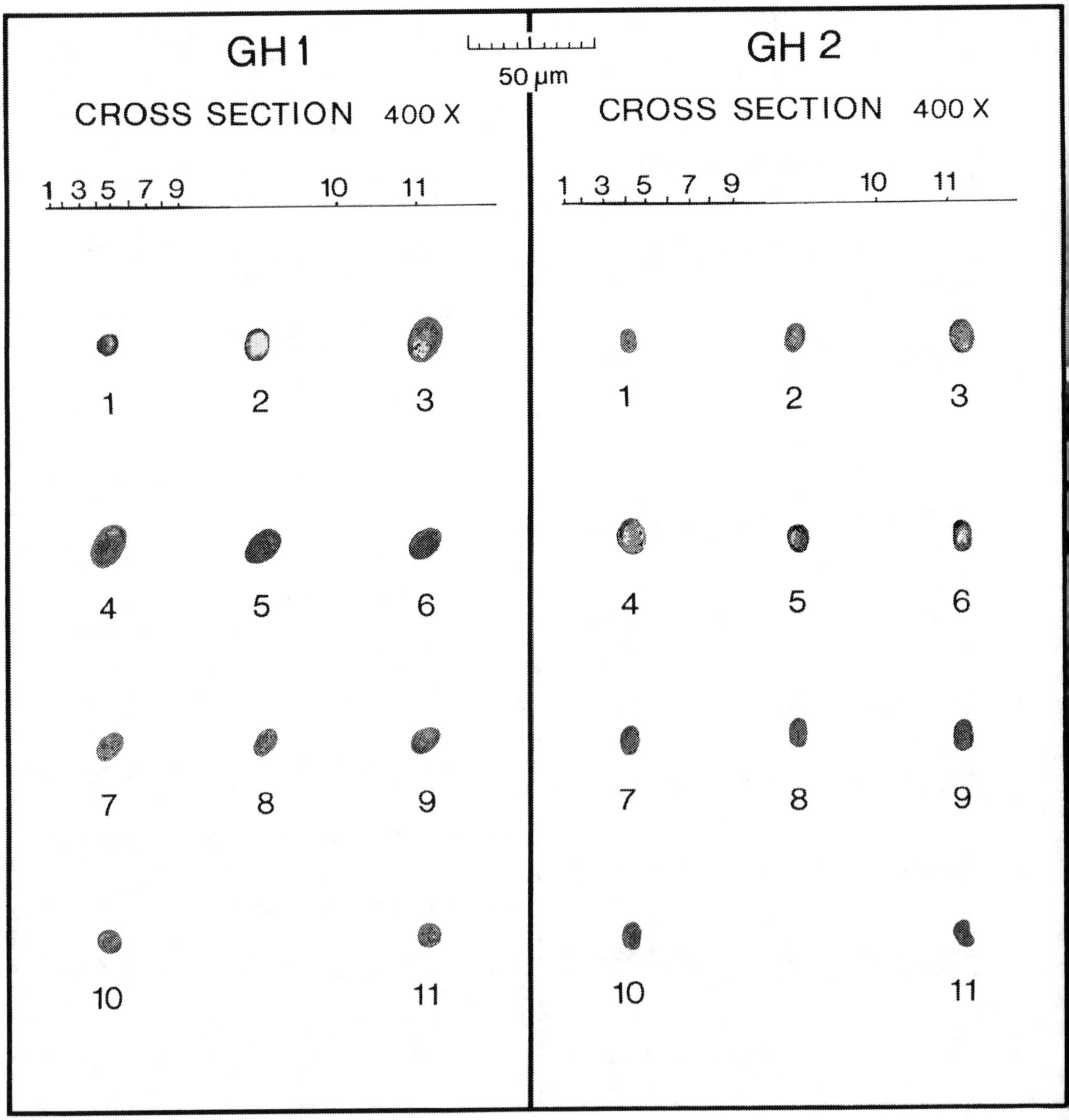

GH 1		GH 2

50 µm

CROSS SECTION   400 X

CROSS SECTION   400 X

1 3 5 7 9   10   11

1 3 5 7 9   10   11

1

2

3

1

2

3

4

5

6

4

5

6

7

8

9

7

8

9

10

11

10

11

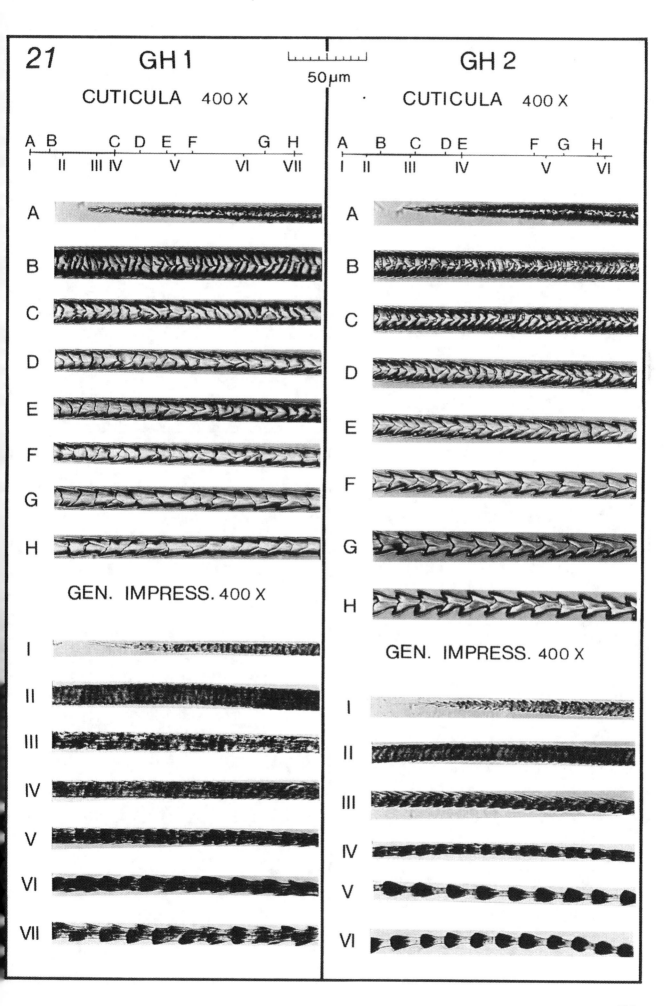

21 GH 1

CUTICULA 400 X

50 µm

GH 2

CUTICULA 400 X

A B   C D E F   G H
I   II   III IV   V   VI   VII

A B C D E   F G H
I   II   III   IV   V   VI

A
B
C
D
E
F
G
H

GEN. IMPRESS. 400 X

A
B
C
D
E
F
G
H

GEN. IMPRESS. 400 X

I
II
III
IV
V
VI
VII

I
II
III
IV
V
VI

22 Nyctalus noctula

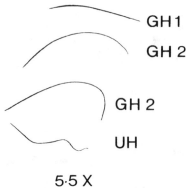

GH 1

GH 2

GH 2

UH

5·5 X

GH 1 | 50 µm | GH 2

CROSS SECTION 400 X

CROSS SECTION 400 X

| 1 | 3 | 5 | 7 | 9 | 10 | 11 |

| 1 | 3 | 5 | 7 | 9 | 10 | 11 |

1 2 3

1 2 3

4 5 6

4 5 6

7 8 9

7 8 9

10 11

10 11

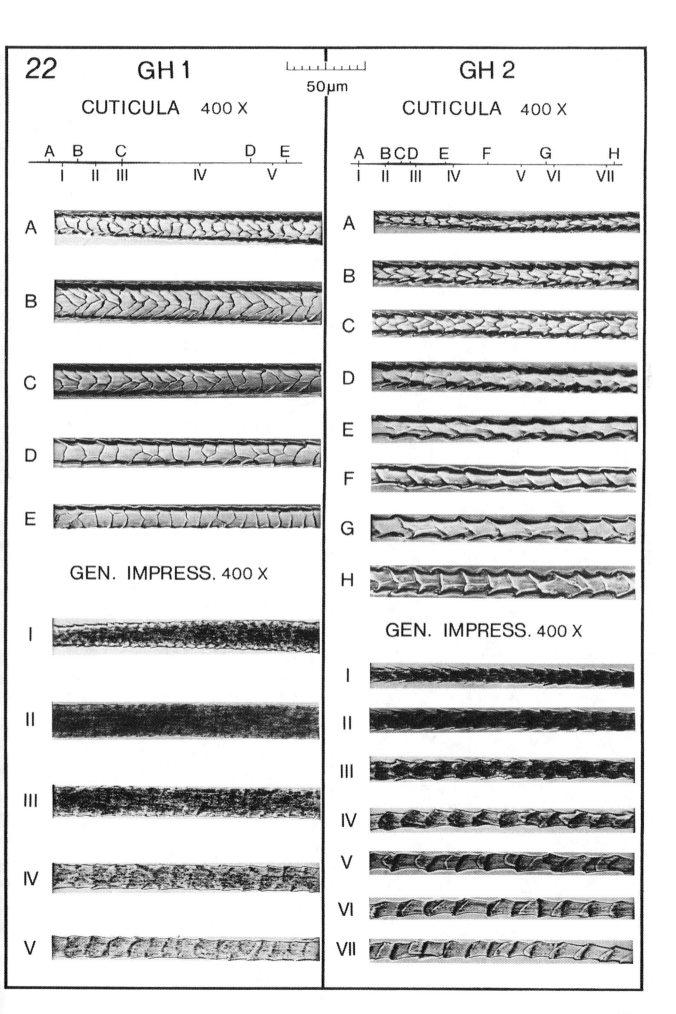

23 *Nyctalus leisleri*

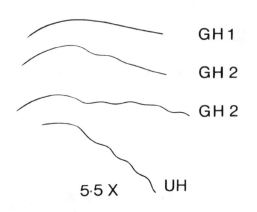

GH 1

GH 2

GH 2

5·5 X   UH

GH 1	50 µm	GH 2

CROSS SECTION 400 X   CROSS SECTION 400 X

1 3 5 7 9   10   11   1 3 5 7 9   10   11

1   2   3   1   2   3

4   5   6   4   5   6

7   8   9   7   8   9

10   11   10   11

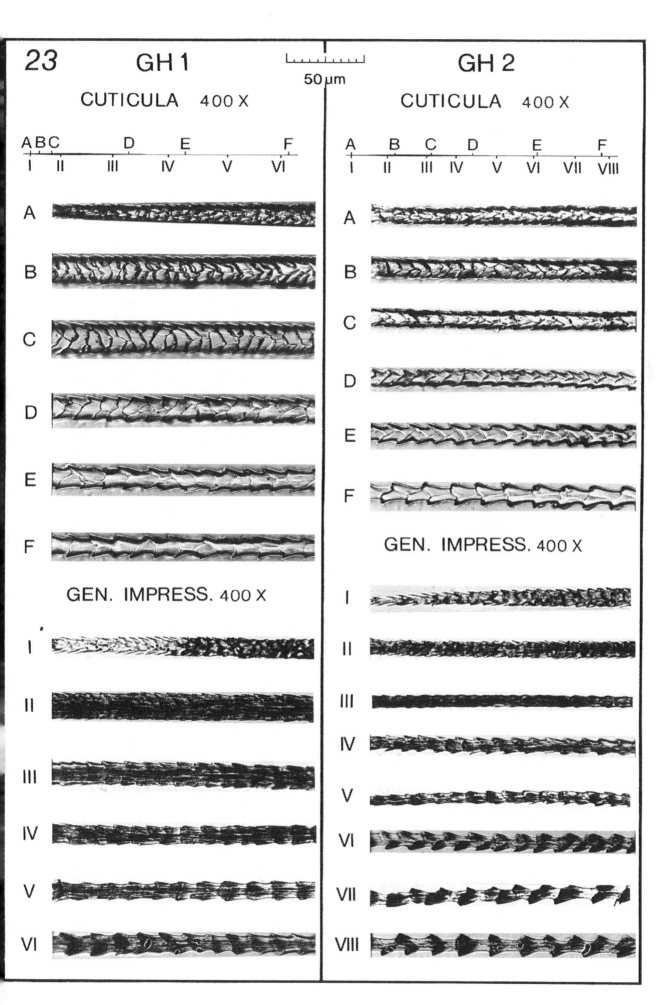

23 GH 1 50 µm GH 2

CUTICULA 400 X   CUTICULA 400 X

GEN. IMPRESS. 400 X

GEN. IMPRESS. 400 X

117

24 Eptesicus serotinus

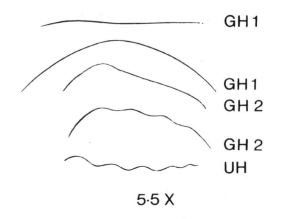

GH 1

GH 1
GH 2

GH 2
UH

5·5 X

GH 1	GH 2
CROSS SECTION 400 X	CROSS SECTION 400 X

50µm

GH 1 — scale: 1, 3, 5, 7, 9, 10, 11

GH 2 — scale: 1, 3, 5, 7, 9, 10, 11

GH 1: 1 2 3 4 5 6 7 8 9 10 11

GH 2: 1 2 3 4 5 6 7 8 9 10 11

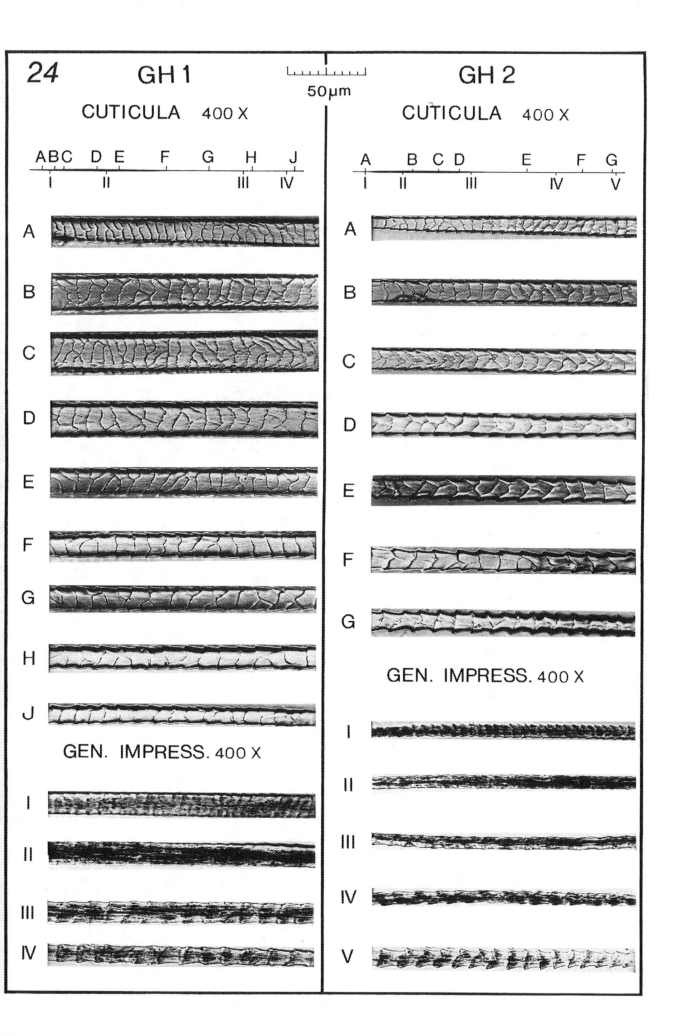

24 GH 1

CUTICULA 400 X

GH 2

CUTICULA 400 X

GEN. IMPRESS. 400 X

GEN. IMPRESS. 400 X

25 Vespertilio murinus

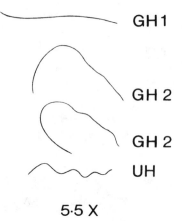

GH 1
GH 2
GH 2
UH

5·5 X

GH 1		GH 2

50μm

CROSS SECTION 400 X   CROSS SECTION 400 X

1 3 5 7 9   10   11   1 3 5 7 9   10   11

1   2   3   1   2   3

4   5   6   4   5   6

7   8   9   7   8   9

10   11   10   11

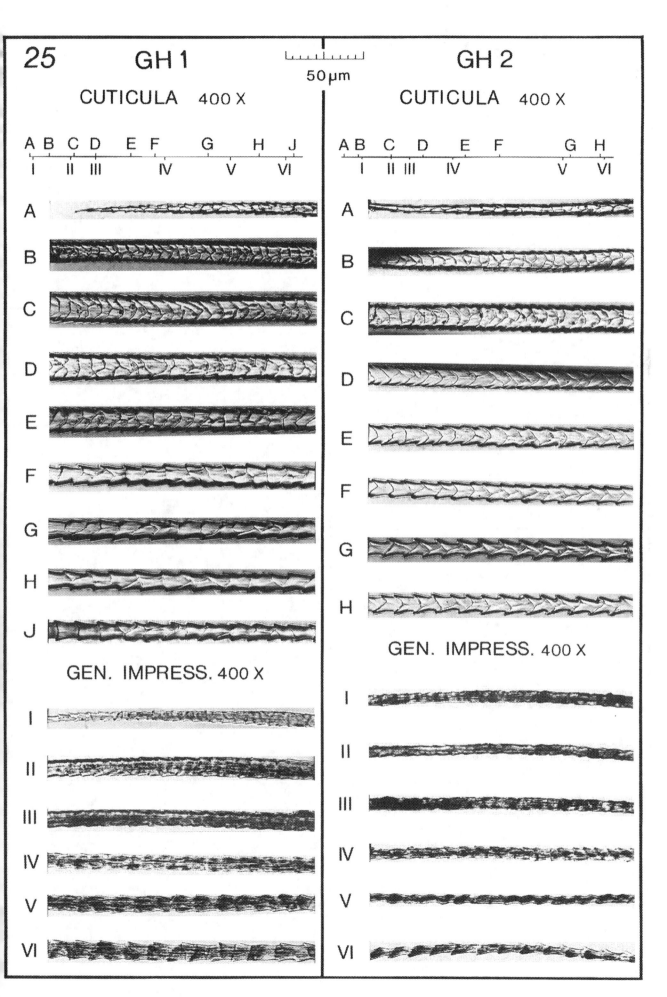

25 GH 1 CUTICULA 400X 50µm GH 2 CUTICULA 400X

GEN. IMPRESS. 400X

26 Barbastella barbastellus

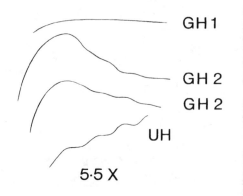

GH 1
GH 2
GH 2
UH

5·5 X

GH 1		GH 2	
CROSS SECTION 400 X	50μm	CROSS SECTION 400 X	

GH 1

1   3   5   7   9   10   11

1   2   3

4   5   6

7   8   9

10   11

GH 2

1   3   5   7   9   10   11

1   2   3

4   5   6

7   8   9

10   11

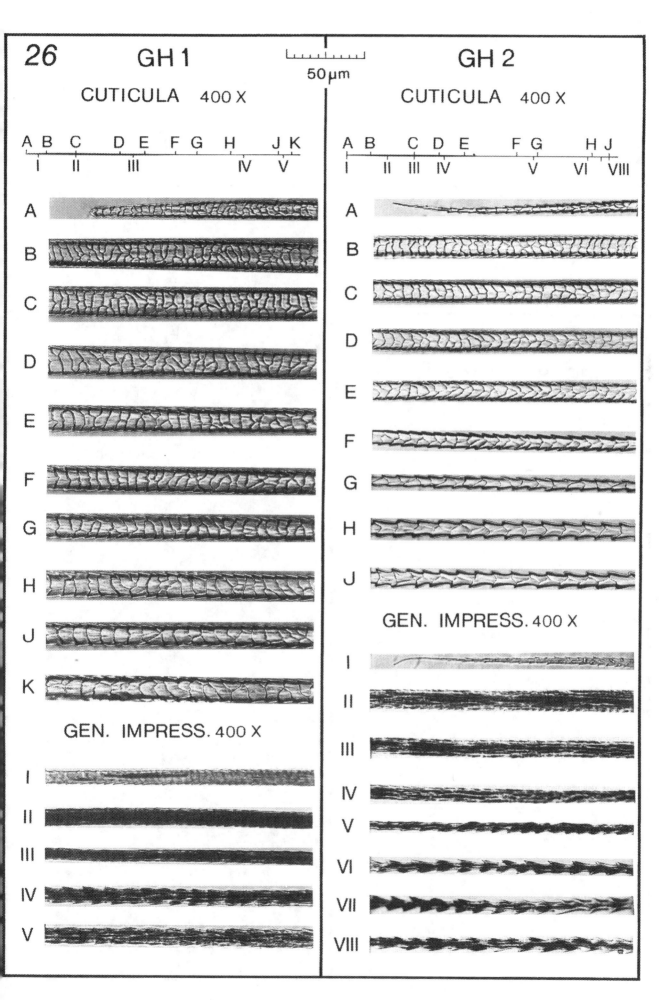

27 *Plecotus auritus*

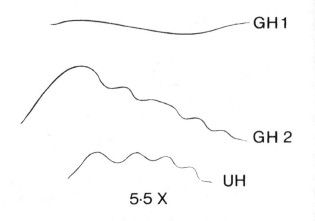

GH1

GH2

UH

5·5 X

GH 1

50µm

GH 2

CROSS SECTION   400 X

CROSS SECTION   400 X

1 3 5 7 9   10   11

1 3 5 7 9   10   11

1

2

3

1

2

3

4

5

6

4

5

6

7

8

9

7

8

9

10

11

10

11

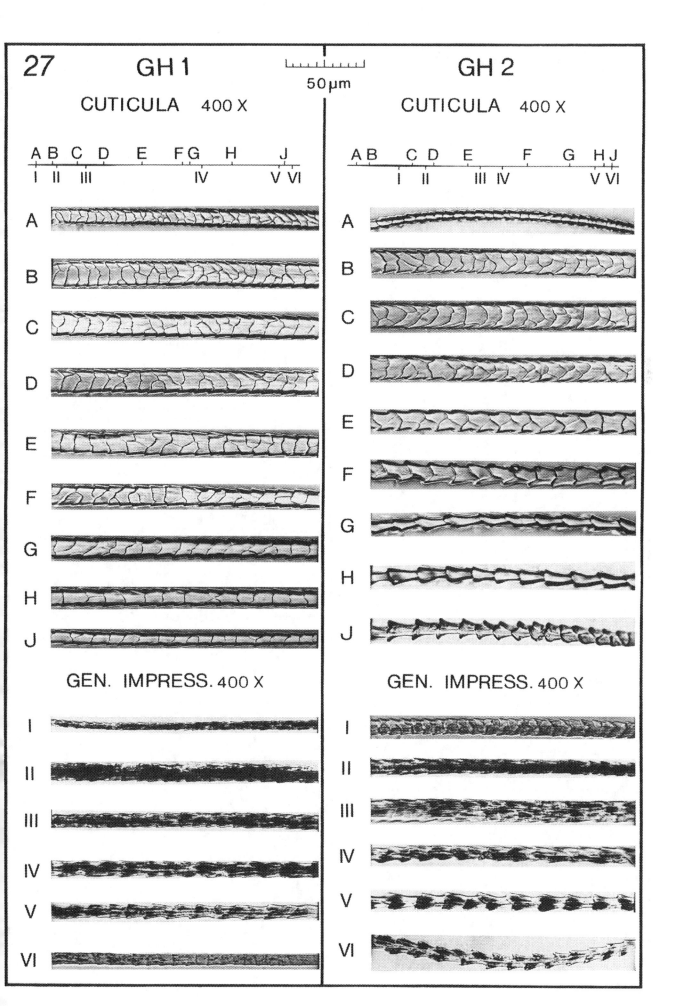

27 GH 1 50µm GH 2

CUTICULA 400 X   CUTICULA 400 X

GEN. IMPRESS. 400 X   GEN. IMPRESS. 400 X

28 Plecotus austriacus

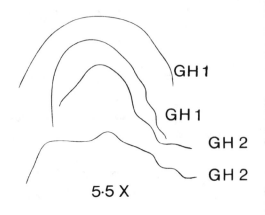

GH 1

GH 1

GH 2

GH 2

5·5 X

| GH 1 | | 50 µm | GH 2 | |

GH 1

CROSS SECTION 400 X

1 3 5 7 9   10   11

1   2   3

4   5   6

7   8   9

10   11

GH 2

CROSS SECTION 400 X

1 3 5 7 9   10   11

1   2   3

4   5   6

7   8   9

10   11

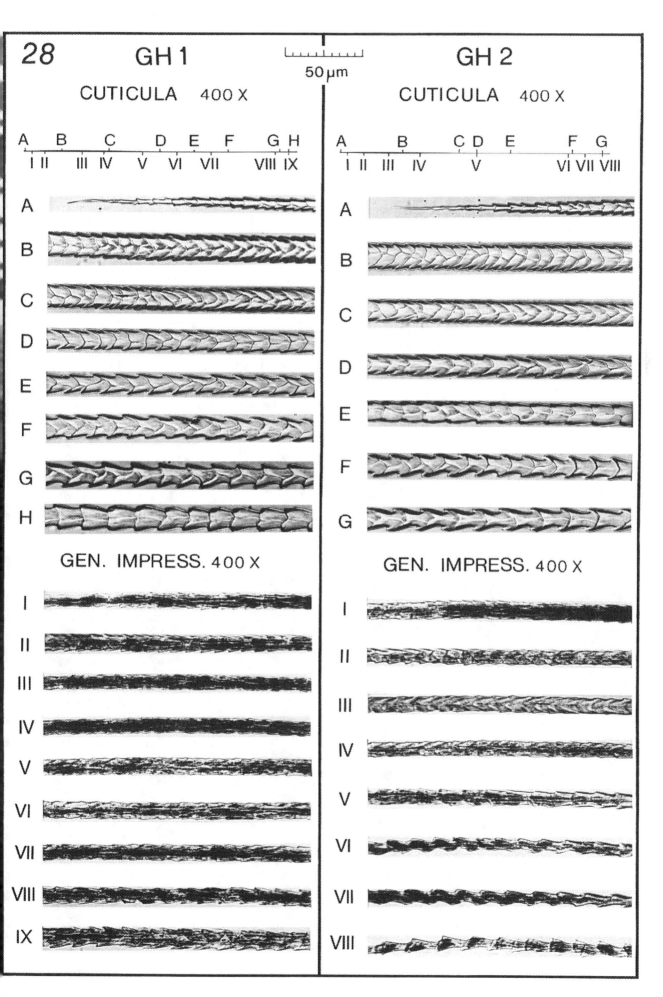

28

GH 1

CUTICULA 400 X

50 µm

A B C D E F G H
I II III IV V VI VII VIII IX

A
B
C
D
E
F
G
H

GEN. IMPRESS. 400 X

I
II
III
IV
V
VI
VII
VIII
IX

GH 2

CUTICULA 400 X

A B C D E F G
I II III IV V VI VII VIII

A
B
C
D
E
F
G

GEN. IMPRESS. 400 X

I
II
III
IV
V
VI
VII
VIII

127

29 *Cricetus cricetus*

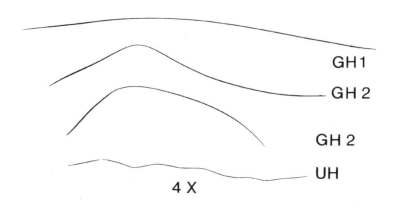

GH1
GH 2
GH 2
UH
4 X

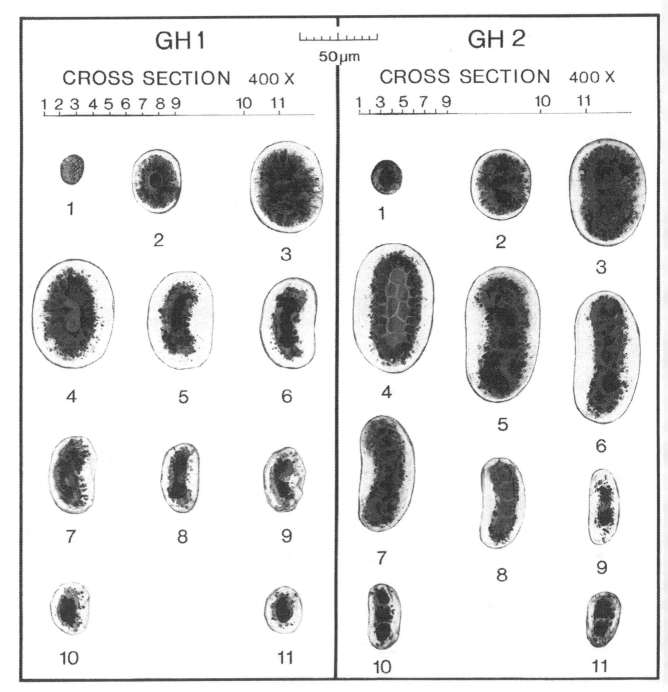

GH1	GH 2
CROSS SECTION 400 X	CROSS SECTION 400 X

50 µm

GH1 scale: 1 2 3 4 5 6 7 8 9 10 11

GH 2 scale: 1 3 5 7 9 10 11

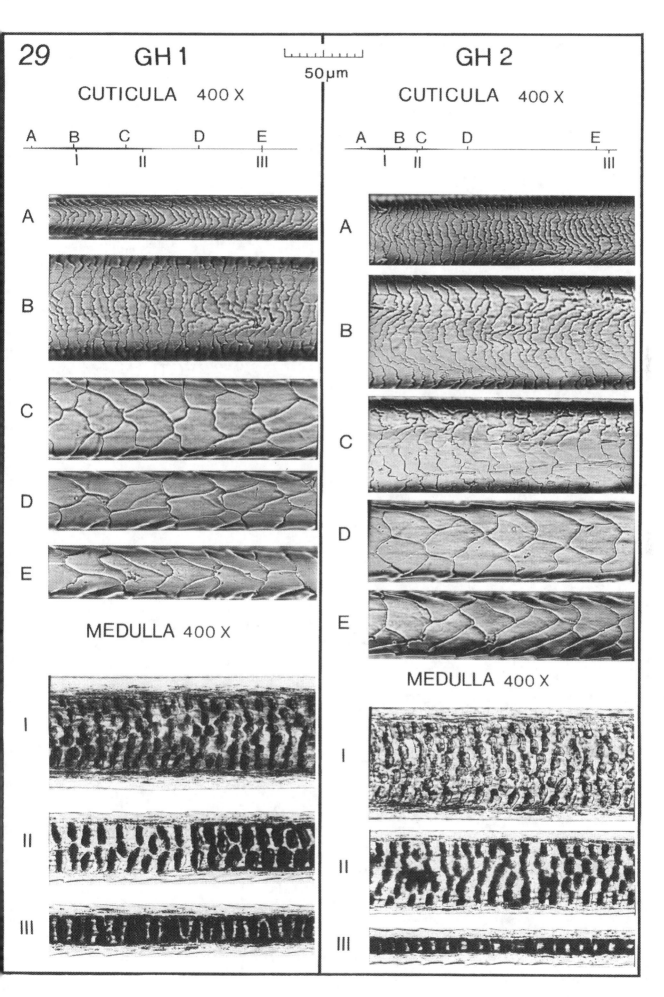

30 Clethrionomys glareolus

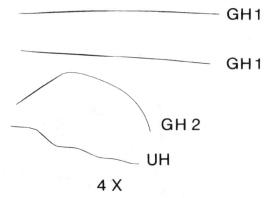

GH 1

GH 1

GH 2

UH

4 X

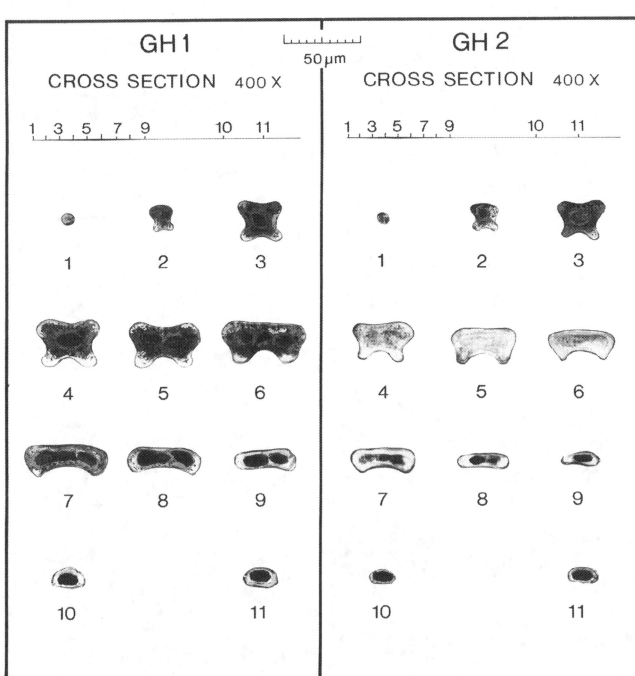

GH 1		GH 2
CROSS SECTION 400 X	50 µm	CROSS SECTION 400 X

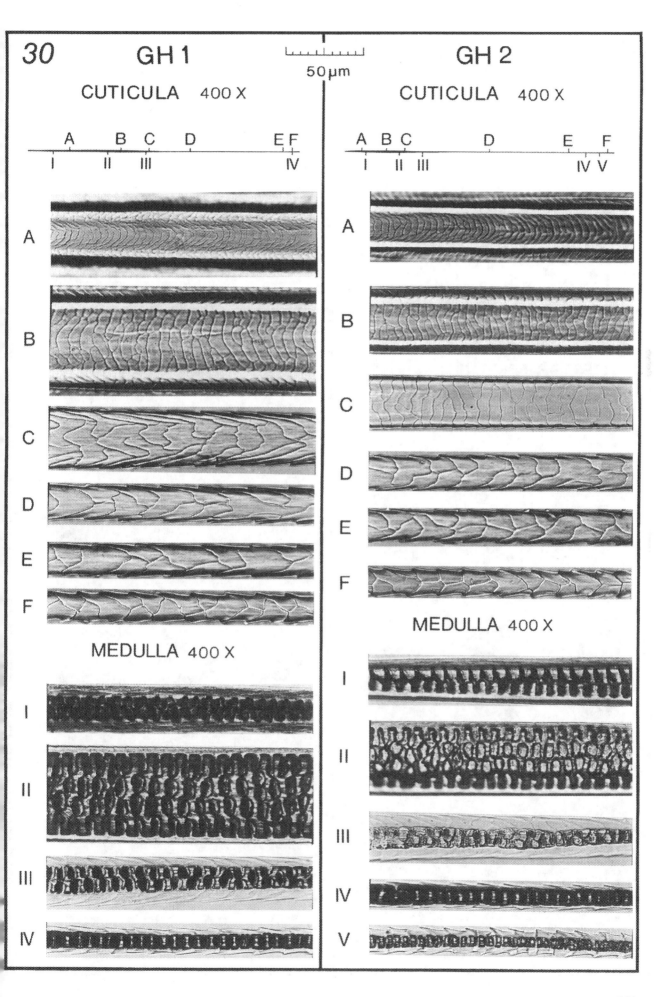

31 Arvicola terrestris

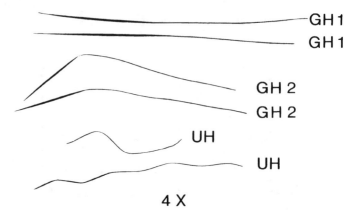

GH 1
GH 1
GH 2
GH 2
UH
UH

4 X

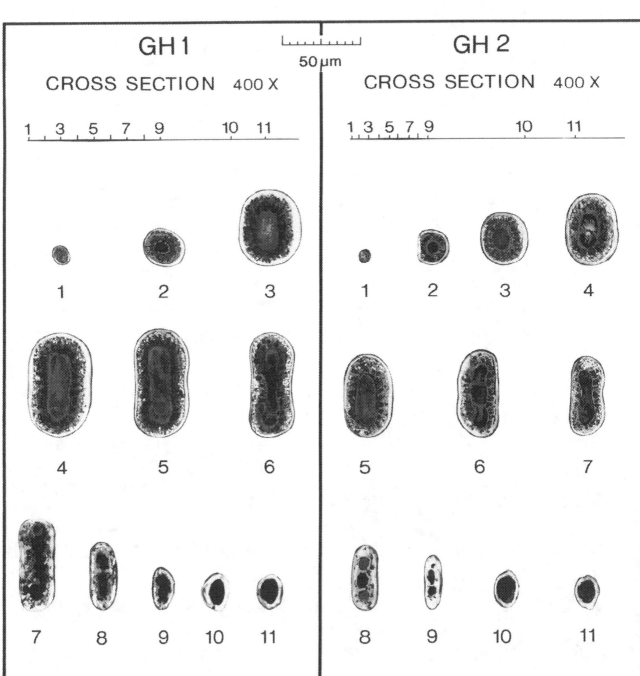

GH 1

CROSS SECTION 400 X

1 3 5 7 9   10 11

1   2   3

4   5   6

7   8   9   10   11

50 µm

GH 2

CROSS SECTION 400 X

1 3 5 7 9   10   11

1   2   3   4

5   6   7

8   9   10   11

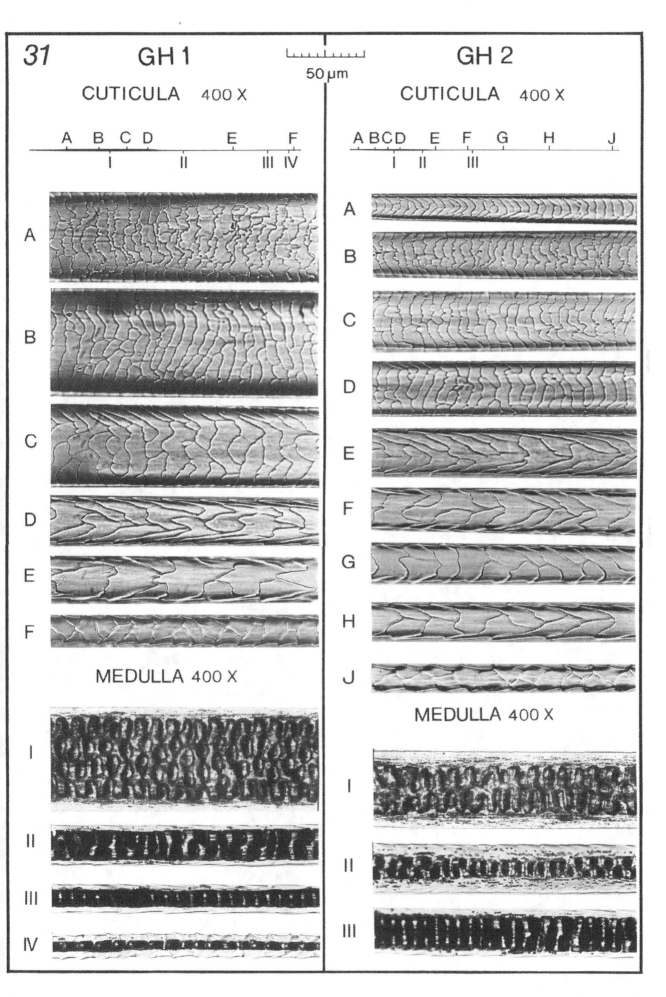

31   GH 1   GH 2

CUTICULA 400 X   CUTICULA 400 X

50 µm

MEDULLA 400 X

MEDULLA 400 X

133

32 Ondatra zibethicus

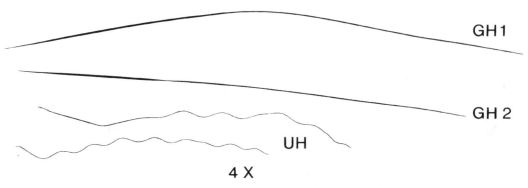

GH 1

GH 2

UH

4 X

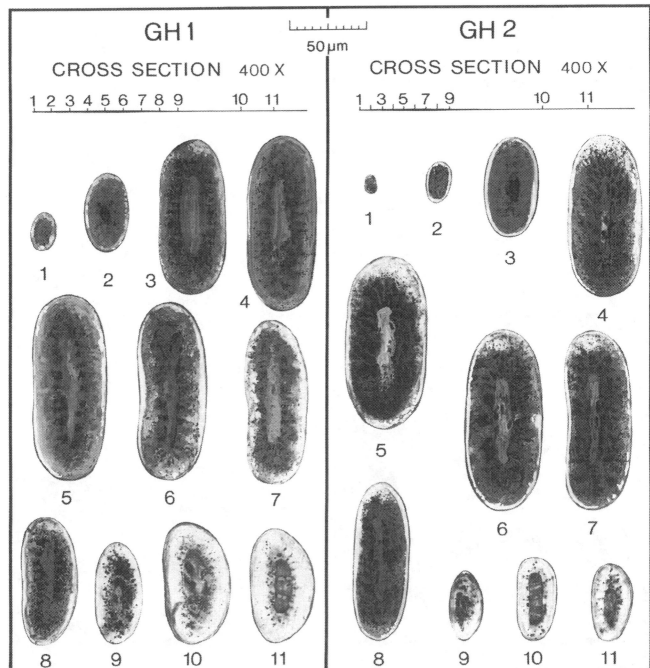

GH 1

50 µm

GH 2

CROSS SECTION 400 X

1 2 3 4 5 6 7 8 9   10   11

CROSS SECTION 400 X

1 3 5 7 9   10   11

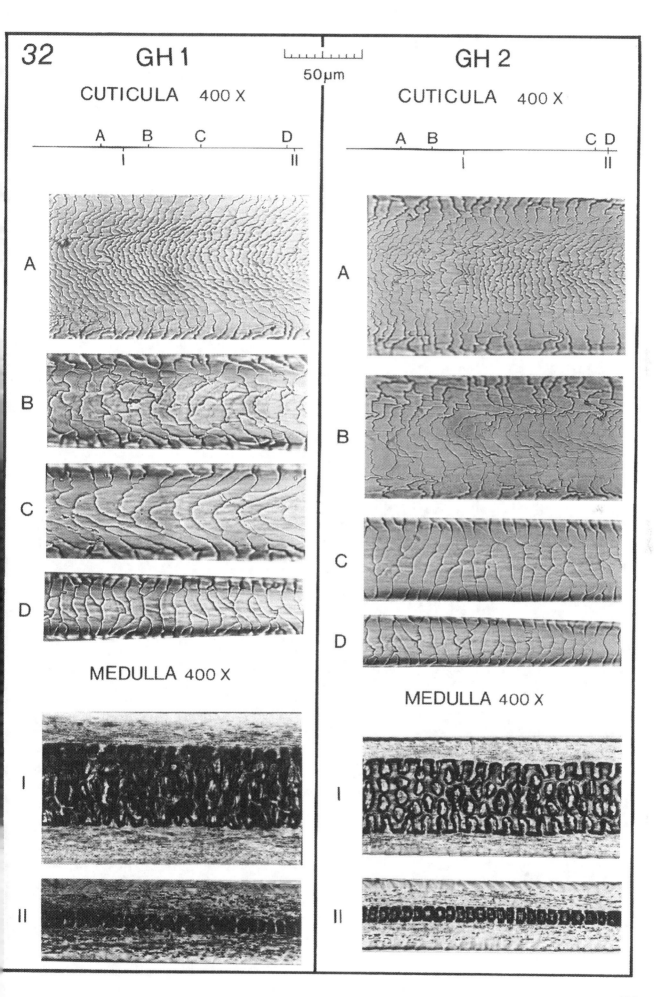

32   GH 1   50µm   GH 2

CUTICULA   400 X   CUTICULA   400 X

A   B   C   D   A B   C D
            I   II   I   II

A   A

B   B

C   C

D   D

MEDULLA 400 X   MEDULLA 400 X

I   I

II   II

33 Pitymys subterraneus

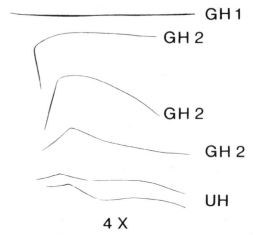

GH 1
GH 2
GH 2
GH 2
UH

4 X

GH 1

CROSS SECTION 400 X

50 µm

GH 2

CROSS SECTION 400 X

1 3 5 7 9   10   11

1 3 5 7 9   10   11

1 2 3

1 2 3

4 5 6 7

4 5 6 7

8 9 10 11

8 9 10 11

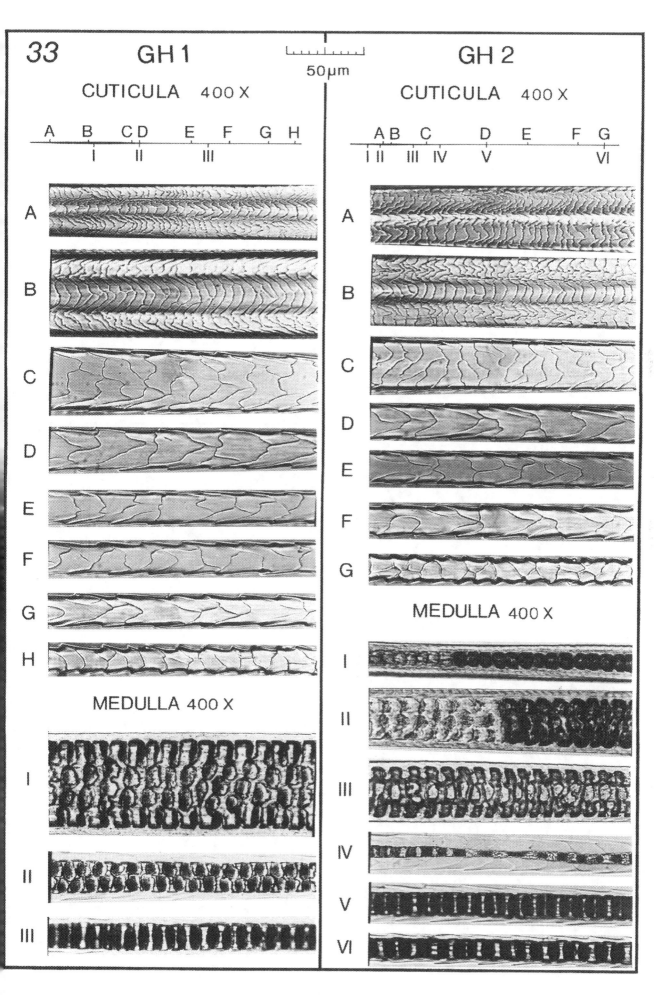

33 GH 1

CUTICULA 400 X

50µm

GH 2

CUTICULA 400 X

MEDULLA 400 X

MEDULLA 400 X

34 Microtus arvalis

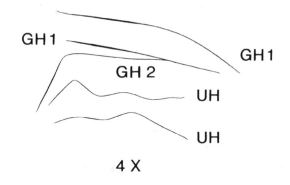

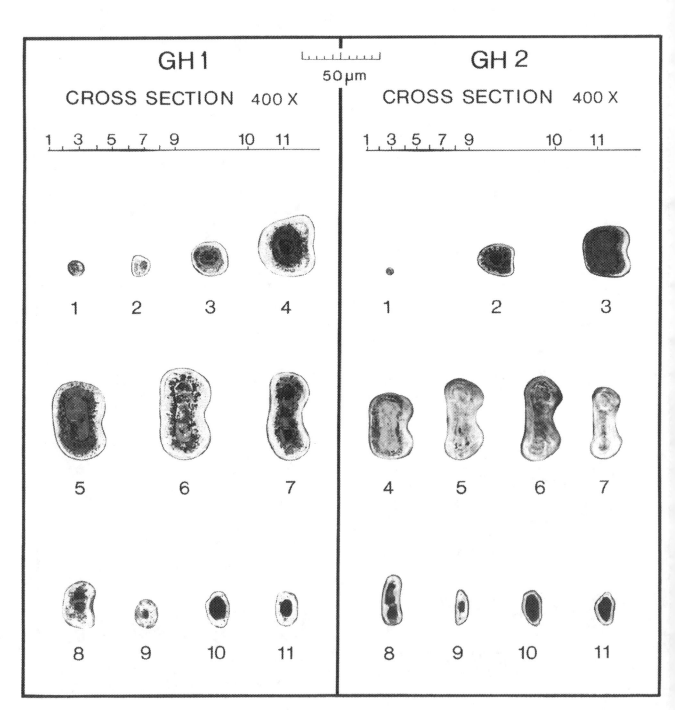

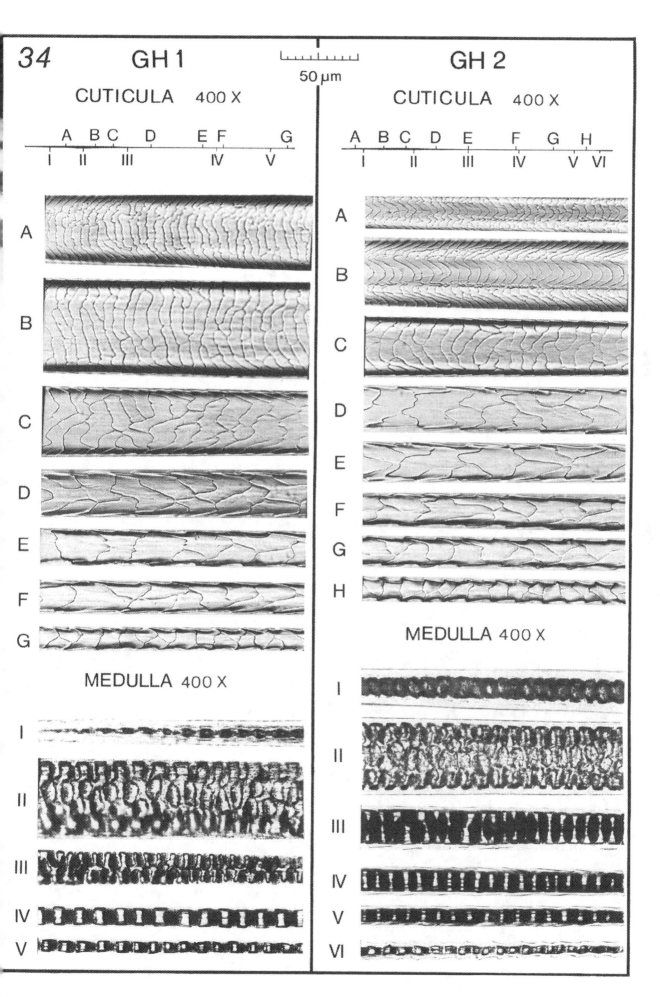

34

GH 1

CUTICULA 400 X

50 µm

A B C D E F G
I II III IV V

A

B

C

D

E

F

G

MEDULLA 400 X

I

II

III

IV

V

GH 2

CUTICULA 400 X

A B C D E F G H
I II III IV V VI

A

B

C

D

E

F

G

H

MEDULLA 400 X

I

II

III

IV

V

VI

35 Microtus agrestis

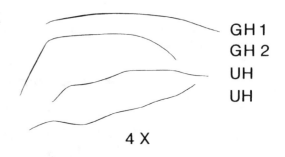

GH 1
GH 2
UH
UH

4 X

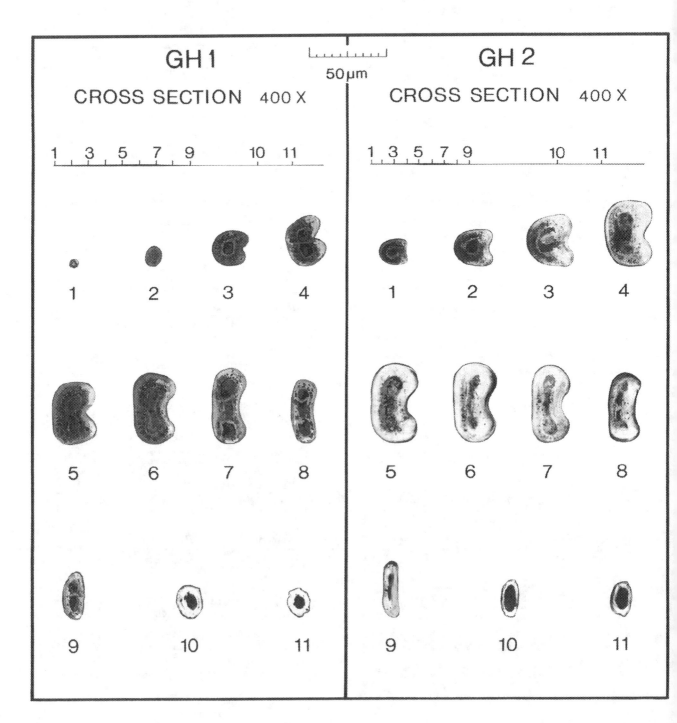

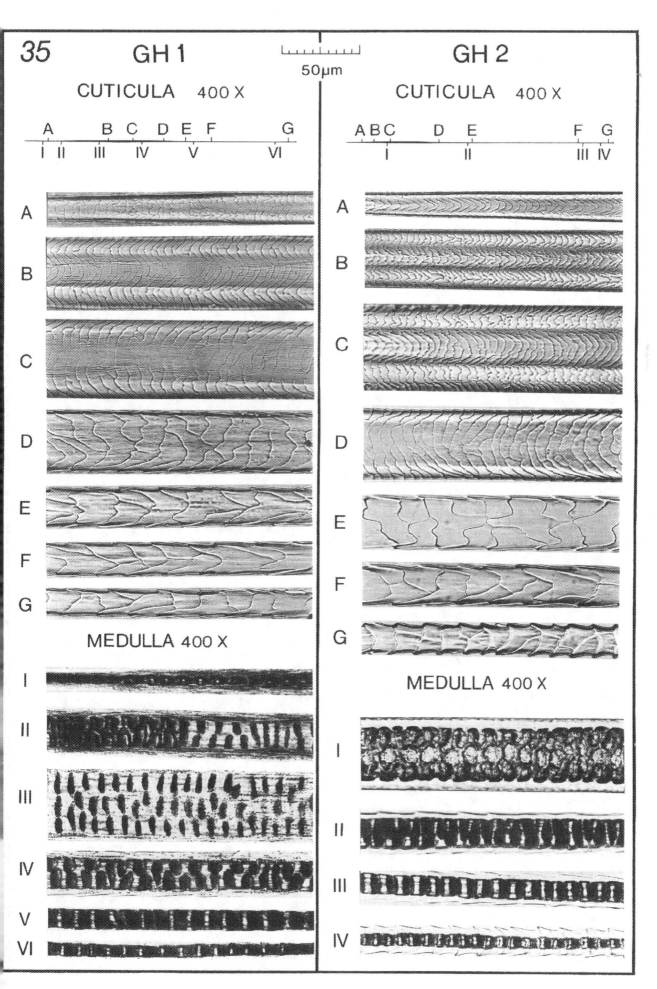

35 GH 1

CUTICULA 400 X

GH 2

CUTICULA 400 X

50µm

MEDULLA 400 X

MEDULLA 400 X

36 Microtus oeconomus

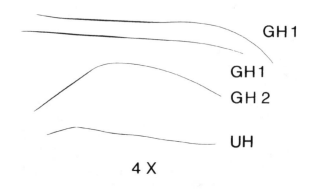

GH 1
GH 1
GH 2
UH

4 X

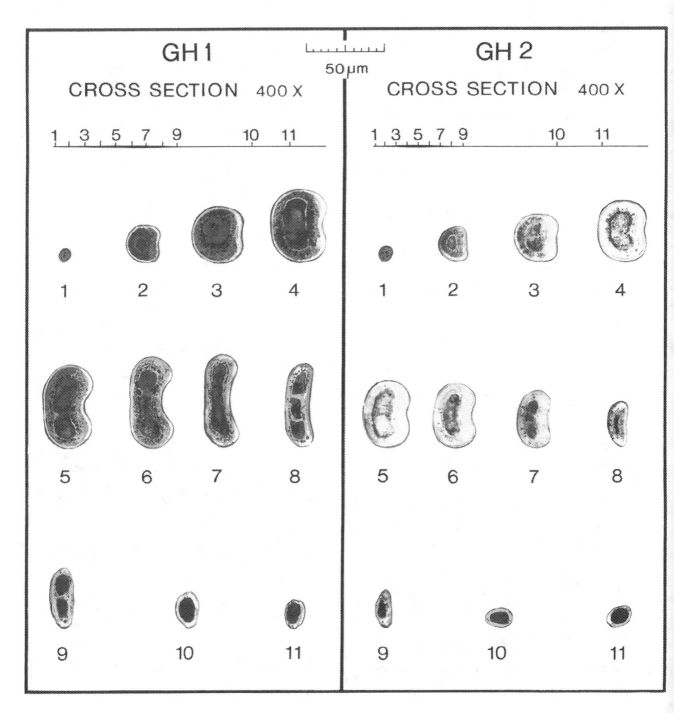

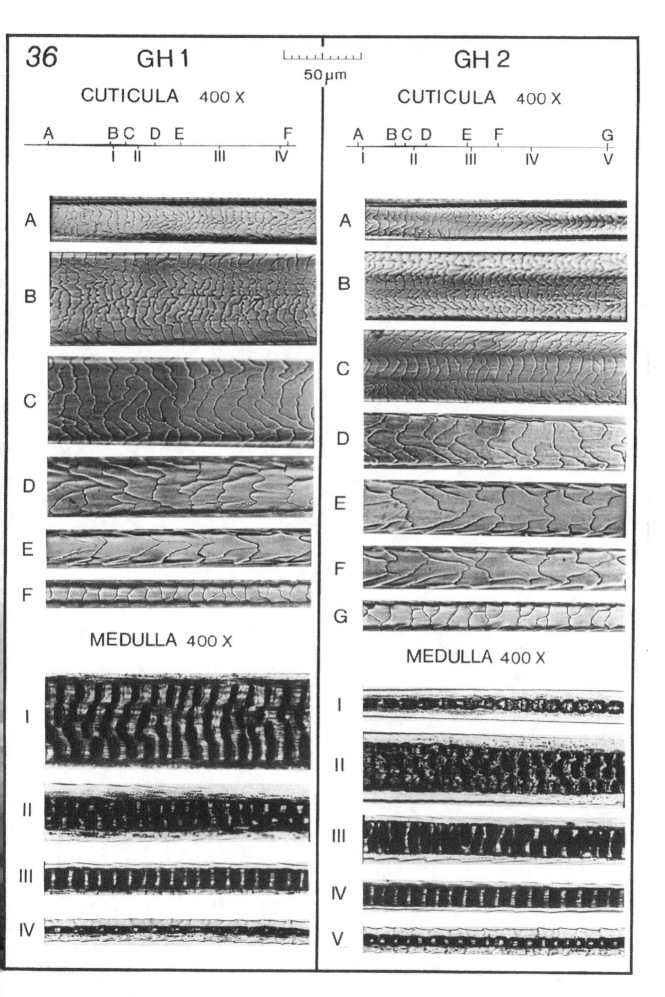

36 GH 1

CUTICULA 400 X

50 µm

A   B C D E   F
          I   II   III   IV

A

B

C

D

E

F

MEDULLA 400 X

I

II

III

IV

GH 2

CUTICULA 400 X

A B C D   E   F   G
   I   II   III   IV   V

A

B

C

D

E

F

G

MEDULLA 400 X

I

II

III

IV

V

37 *Micromys minutus*

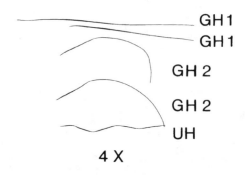

GH 1
GH 1
GH 2
GH 2
UH

4 X

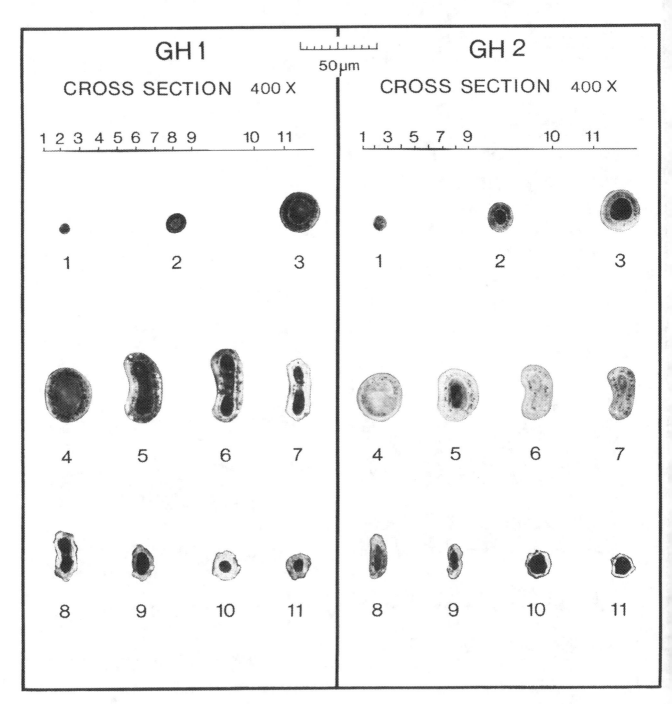

GH 1

GH 2

CROSS SECTION   400 X

CROSS SECTION   400 X

50 µm

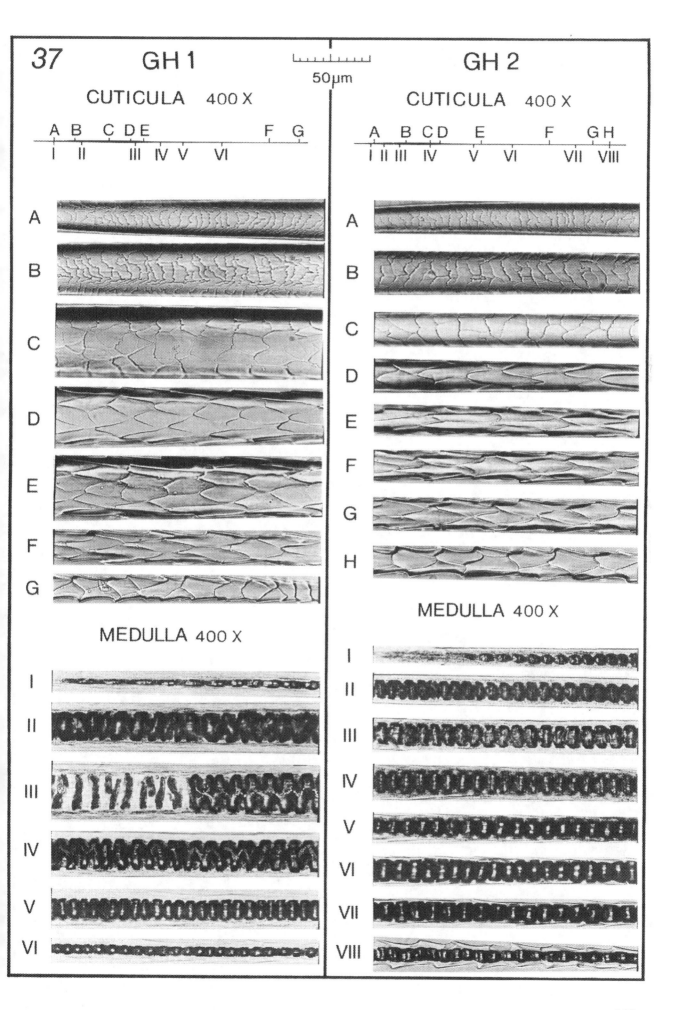

145

38 Apodemus sylvaticus

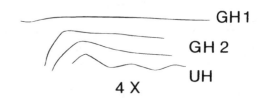

GH 1
GH 2
UH
4 X

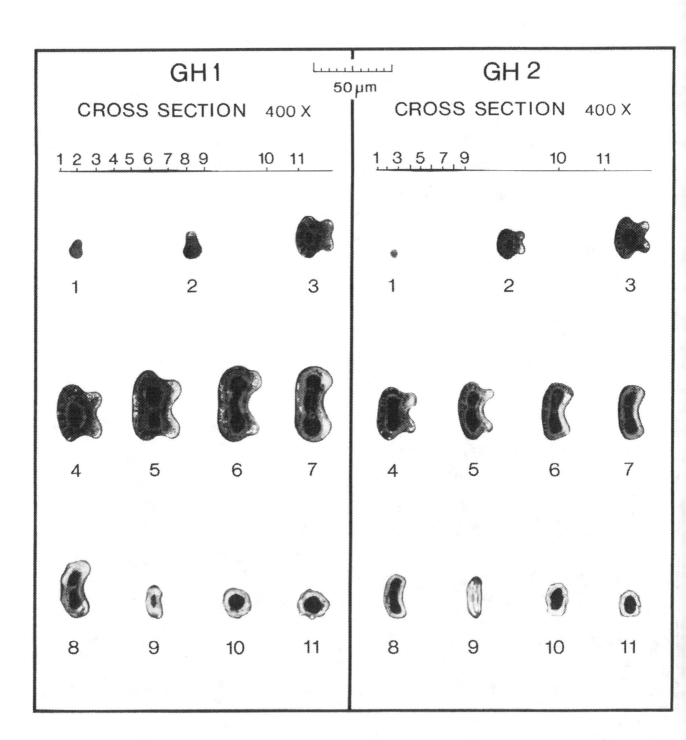

GH 1

CROSS SECTION 400 X

50 μm

GH 2

CROSS SECTION 400 X

1 2 3 4 5 6 7 8 9   10   11

1 3 5 7 9   10   11

1   2   3

1   2   3

4   5   6   7

4   5   6   7

8   9   10   11

8   9   10   11

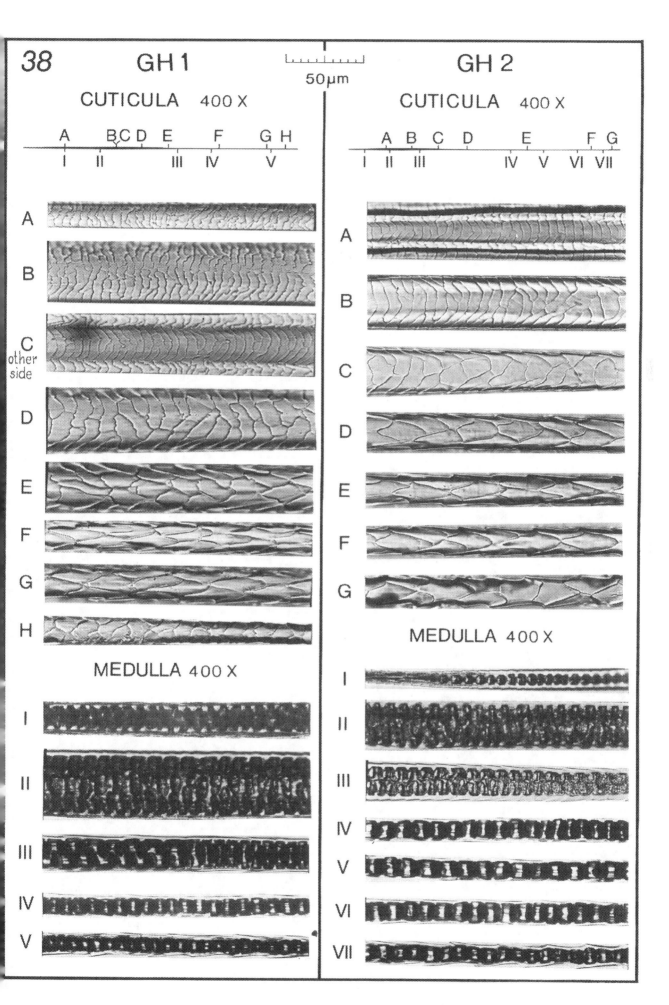

38 GH 1 CUTICULA 400 X

50μm

GH 2 CUTICULA 400 X

MEDULLA 400 X

MEDULLA 400 X

147

39 Apodemus flavicollis

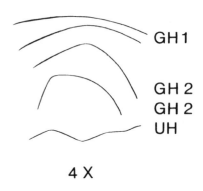

GH1
GH 2
GH 2
UH

4 X

GH 1	50 µm	GH 2
CROSS SECTION 400 X		CROSS SECTION 400 X

1 2 3 4 5 6 7 8 9   10   11

1 3 5 7 9   10   11

1   2   3   1   2   3

4   5   6   7   4   5   6   7

8   9   10   11   8   9   10   11

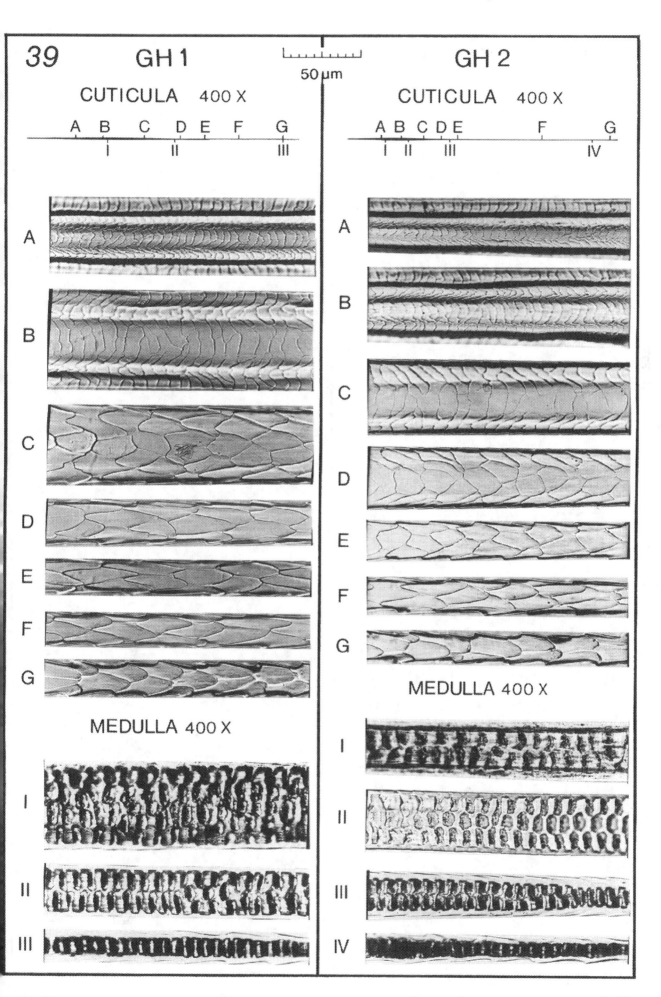

39   GH 1

CUTICULA   400 X

A B C D E F   G
     I   II   III

A

B

C

D

E

F

G

MEDULLA 400 X

I

II

III

GH 2

CUTICULA   400 X

A B C D E   F   G
  I II III   IV

A

B

C

D

E

F

G

MEDULLA 400 X

I

II

III

IV

50 µm

149

40 *Apodemus agrarius*

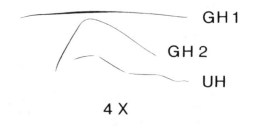

GH 1

GH 2

UH

4 X

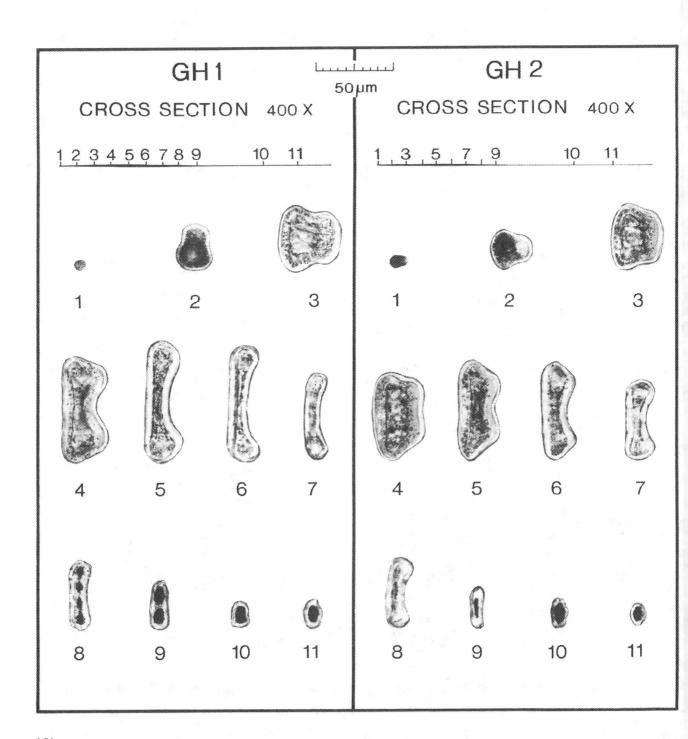

GH 1

50 µm

GH 2

CROSS SECTION   400 X

CROSS SECTION   400 X

1 2 3 4 5 6 7 8 9   10   11

1   3   5   7   9   10   11

1   2   3

1   2   3

4   5   6   7

4   5   6   7

8   9   10   11

8   9   10   11

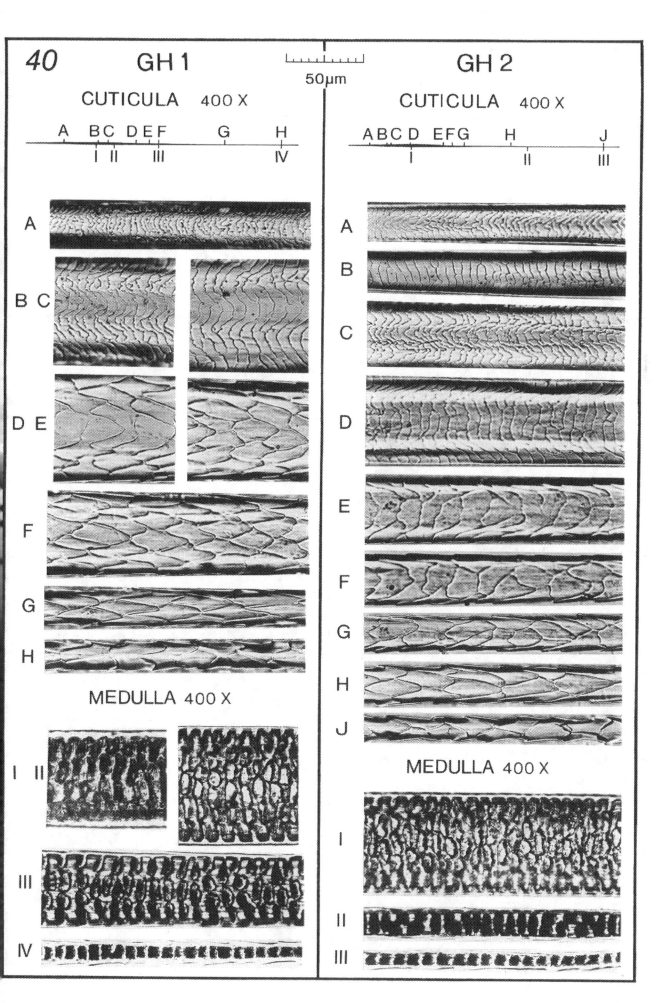

40 GH 1

CUTICULA 400 X

50µm

A BC DEF G H
I II III IV

A

B C

D E

F

G

H

MEDULLA 400 X

I II

III

IV

GH 2

CUTICULA 400 X

ABCD EFG H J
I II III

A

B

C

D

E

F

G

H

J

MEDULLA 400 X

I

II

III

41 Rattus norvegicus

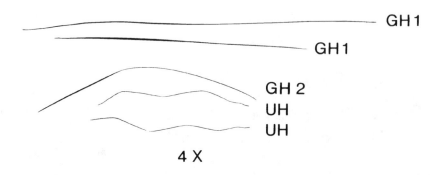

GH 1
GH 1
GH 2
UH
UH

4 X

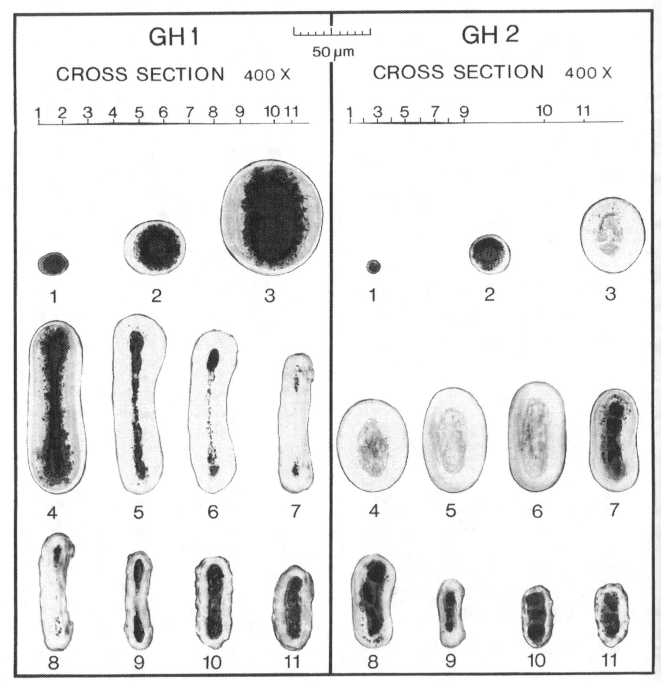

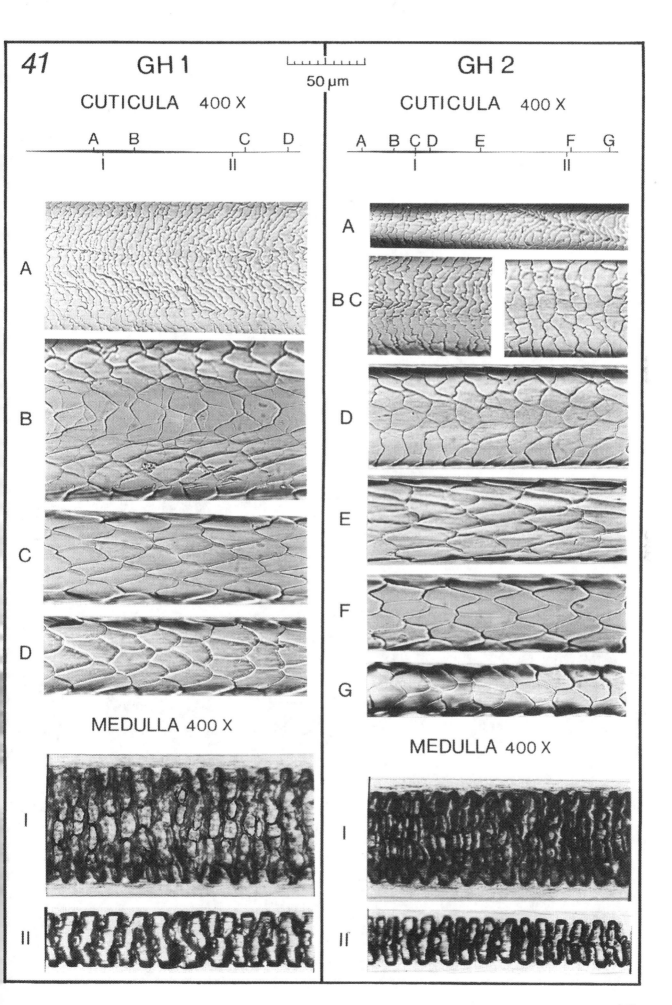

41 GH 1   CUTICULA 400 X

50 µm

A   B   C   D

I   II

A

B

C

D

MEDULLA 400 X

I

II

GH 2   CUTICULA 400 X

A B C D   E   F   G

I   II

A

B C

D

E

F

G

MEDULLA 400 X

I

II

42 Rattus rattus

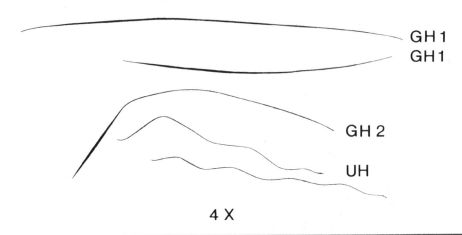

GH 1
GH 1
GH 2
UH

4 X

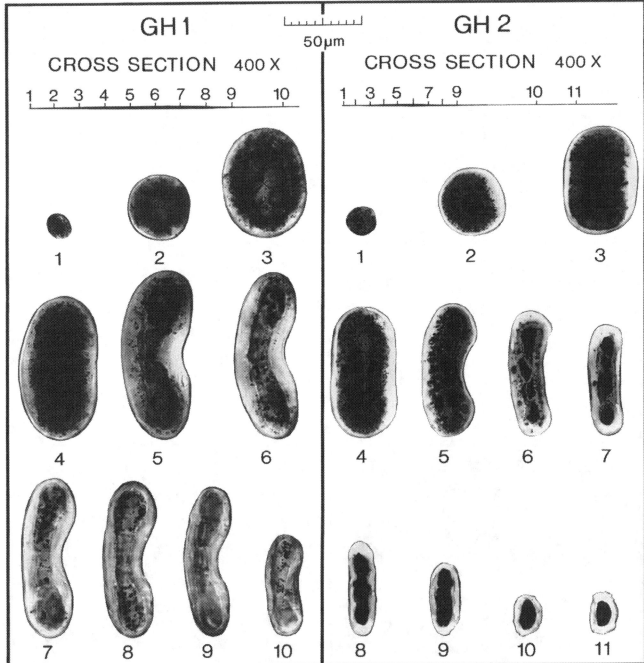

GH 1

CROSS SECTION   400 X

50 µm

GH 2

CROSS SECTION   400 X

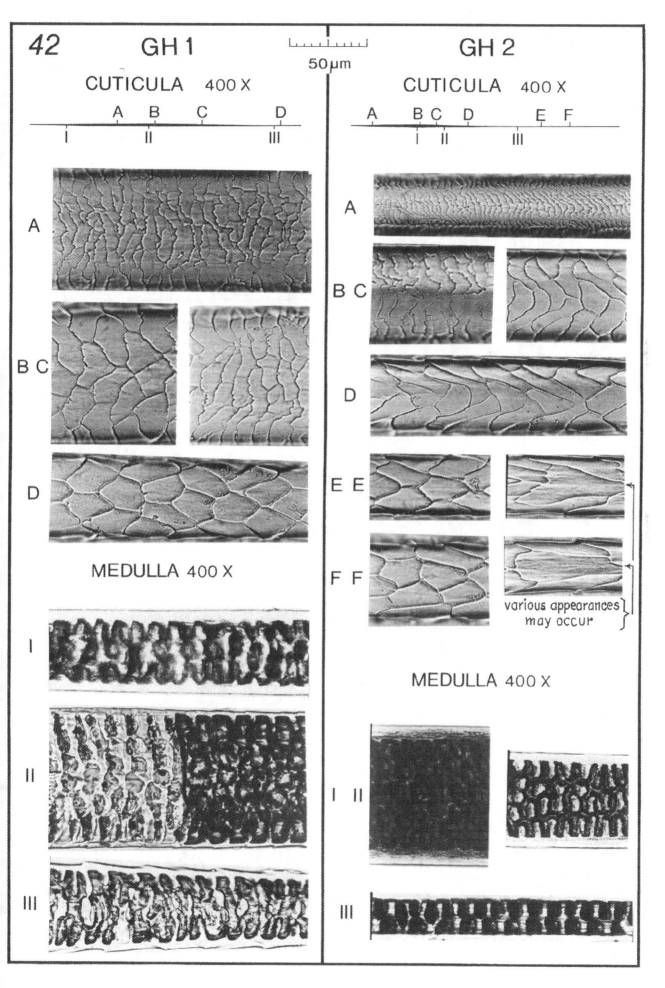

42 GH 1

CUTICULA 400 X

GH 2

CUTICULA 400 X

MEDULLA 400 X

various appearances
may occur

MEDULLA 400 X

50 µm

43 Mus musculus

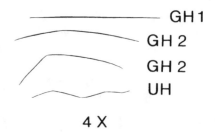

GH 1
GH 2
GH 2
UH

4 X

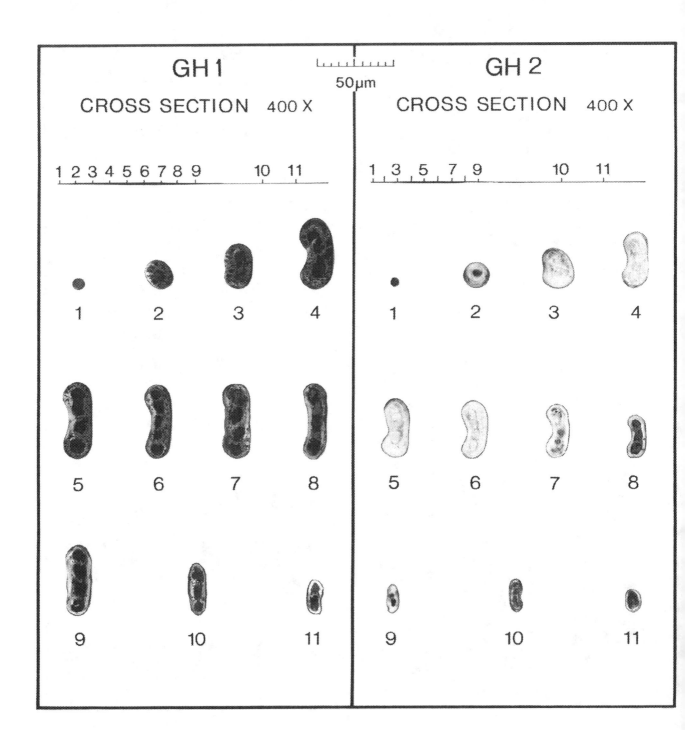

GH 1

CROSS SECTION 400 X

50 µm

GH 2

CROSS SECTION 400 X

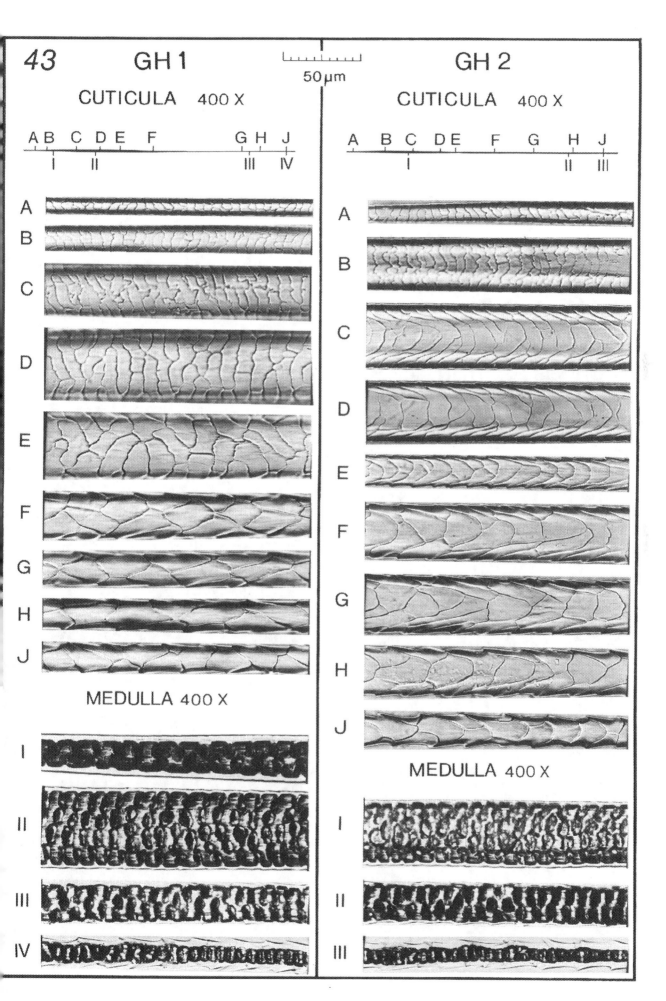

43   GH 1   50 µm   GH 2

CUTICULA   400 X   CUTICULA   400 X

A B C D E   F   G H J   A B C DE   F   G   H   J
     I   II   III IV   I   II   III

A
B
C
D
E
F
G
H
J

MEDULLA 400 X

I
II
III
IV

A
B
C
D
E
F
G
H
J

MEDULLA 400 X

I
II
III

44 Glis glis

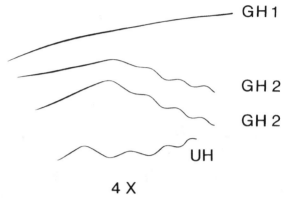

GH 1

GH 2

GH 2

UH

4 X

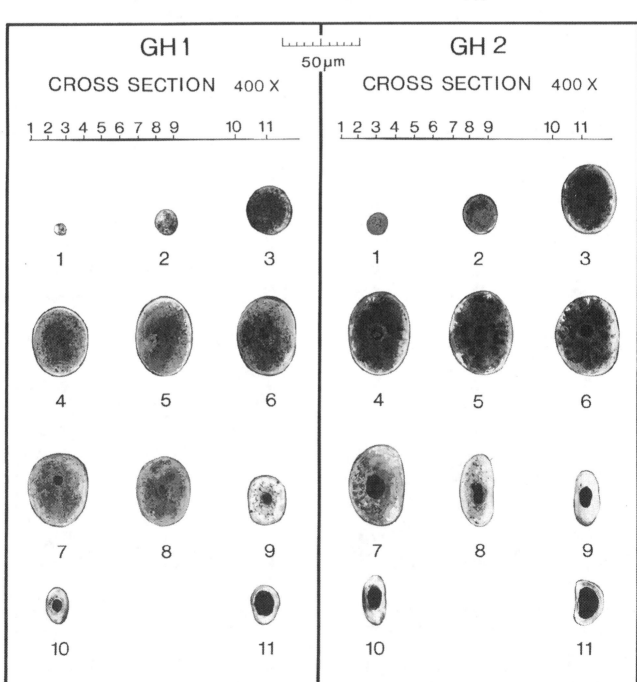

GH 1

CROSS SECTION 400 X

50 µm

GH 2

CROSS SECTION 400 X

1 2 3 4 5 6 7 8 9   10 11

1 2 3 4 5 6 7 8 9   10 11

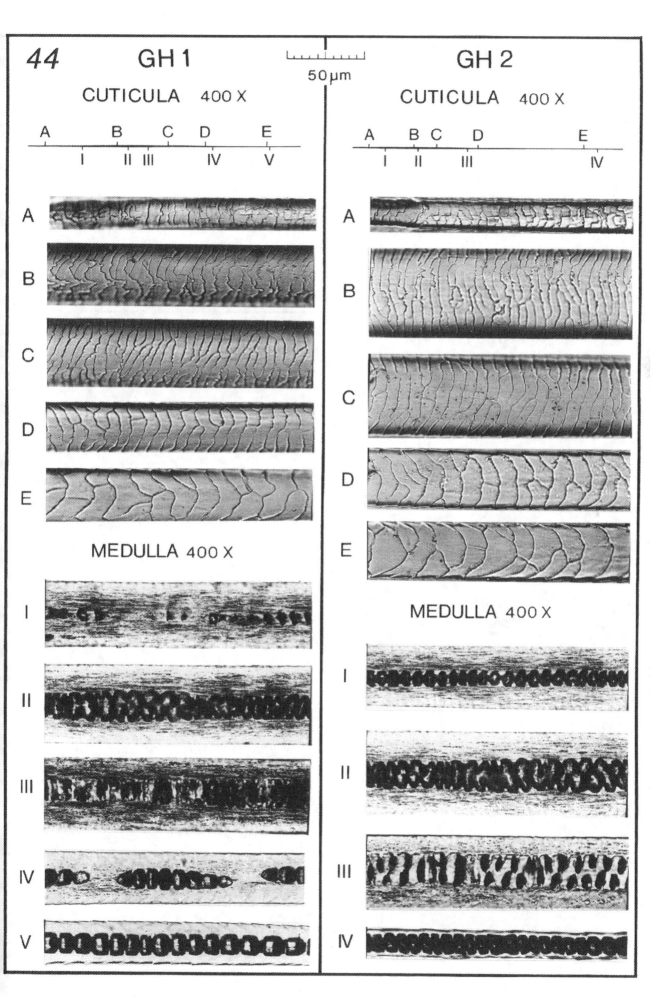

44   GH 1
CUTICULA   400 X

50 µm

A   B   C   D   E
    I   II III   IV   V

A

B

C

D

E

MEDULLA 400 X

I

II

III

IV

V

GH 2
CUTICULA   400 X

A   B C   D   E
    I II   III   IV

A

B

C

D

E

MEDULLA 400 X

I

II

III

IV

45 Muscardinus avellanarius

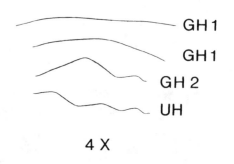

GH 1
GH 1
GH 2
UH

4 X

GH 1

CROSS SECTION   400 X

1 2 3 4 5 6 7 8 9   10 11

1

2

3

4

5

6

7

8

9

10

11

50 µm

GH 2

CROSS SECTION   400 X

1 3 5 7 9   10 11

1

2

3

4

5

6

7

8

9

10

11

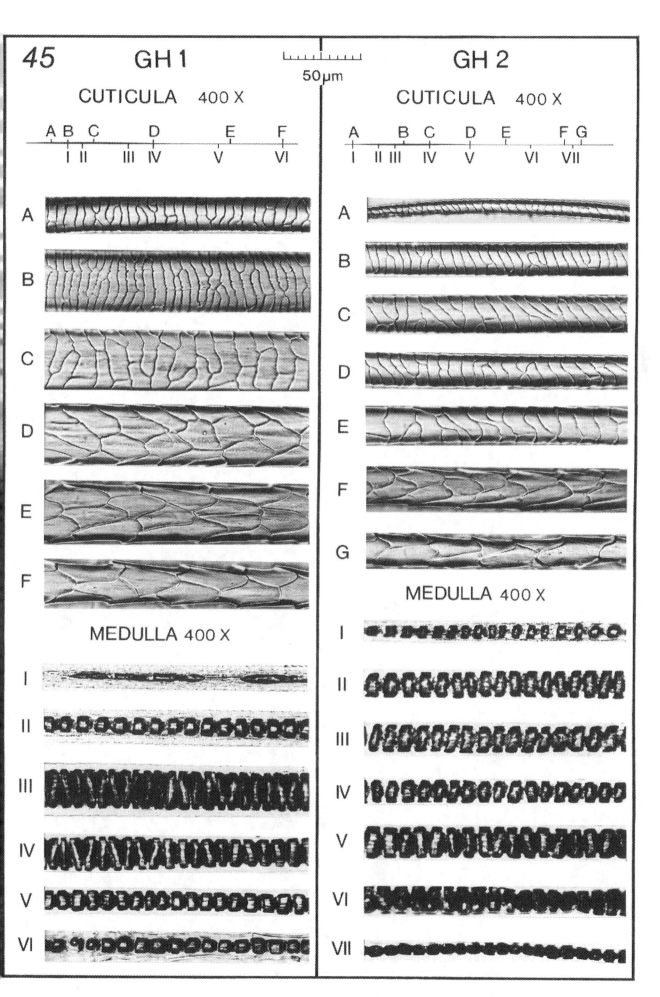

45 GH 1   50µm   GH 2

CUTICULA 400 X

CUTICULA 400 X

MEDULLA 400 X

MEDULLA 400 X

161

46 Eliomys quercinus

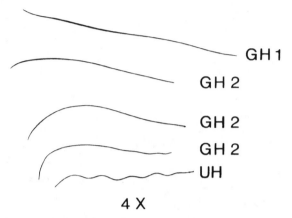

GH 1
GH 2
GH 2
GH 2
UH

4 X

GH 1

50 µm

GH 2

CROSS SECTION 400 X

CROSS SECTION 400 X

1 3 5 7 9   10   11

1 3 5 7 9   10   11

1   2   3

1   2   3

4   5   6

4   5   6

7   8   9

7   8   9

10   11

10   11

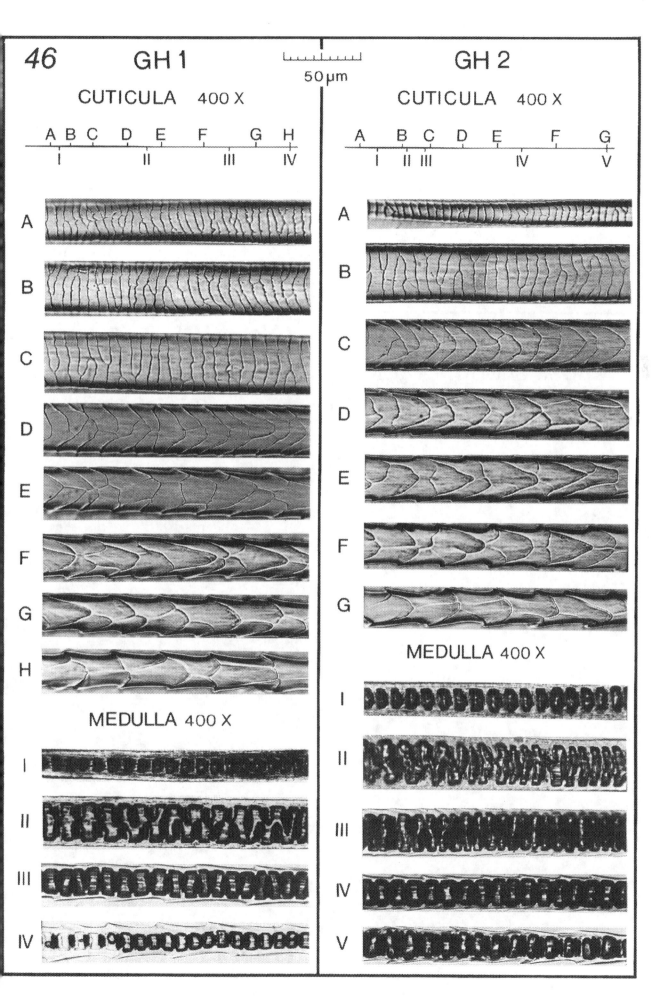

46 GH 1

CUTICULA 400 X

GH 2

CUTICULA 400 X

MEDULLA 400 X

MEDULLA 400 X

47 Castor fiber

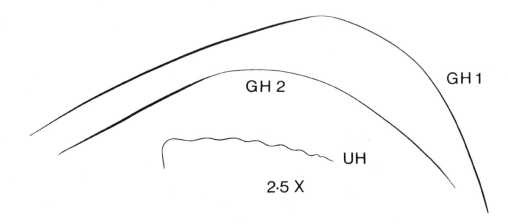

GH 1

GH 2

UH

2·5 X

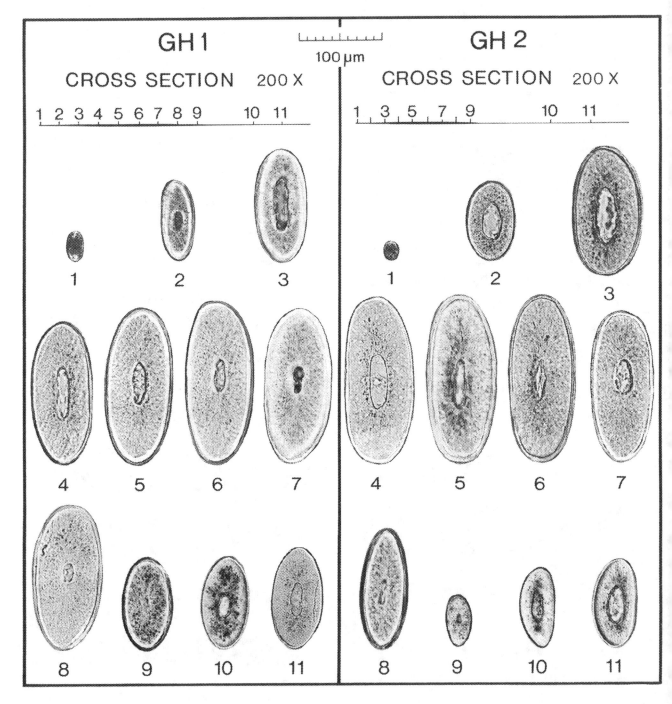

GH 1

100 µm

GH 2

CROSS SECTION   200 X

CROSS SECTION   200 X

1 2 3 4 5 6 7 8 9   10 11

1 3 5 7 9   10   11

1   2   3

1   2   3

4   5   6   7

4   5   6   7

8   9   10   11

8   9   10   11

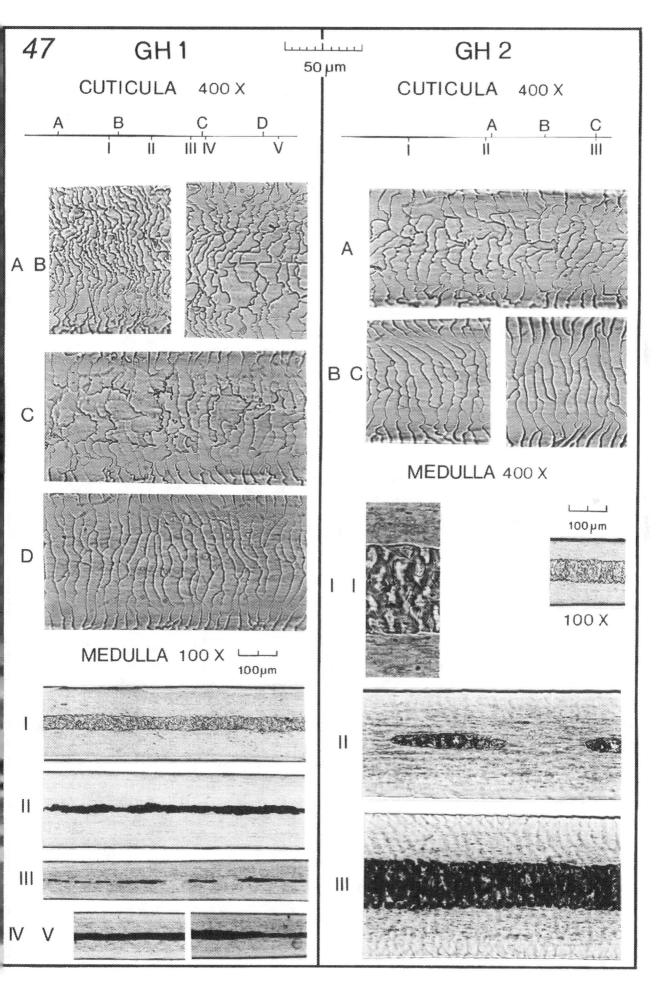

47 GH 1

CUTICULA 400 X

50 μm

A B

C

D

MEDULLA 100 X

100μm

I

II

III

IV V

GH 2

CUTICULA 400 X

A

B C

MEDULLA 400 X

100 μm

I I

100 X

II

III

165

48 Myocastor coypus

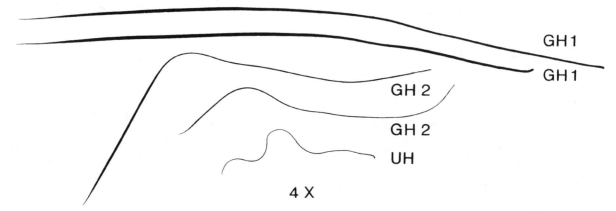

GH 1

GH 1

GH 2

GH 2

UH

4 X

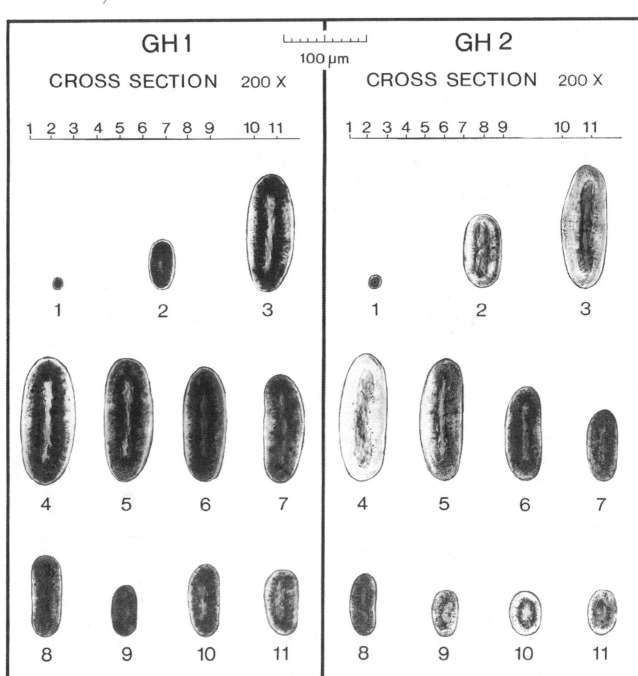

| GH 1 | | 100 μm | GH 2 | |
| CROSS SECTION | 200 X | | CROSS SECTION | 200 X |

1 2 3 4 5 6 7 8 9 10 11

1 2 3 4 5 6 7 8 9 10 11

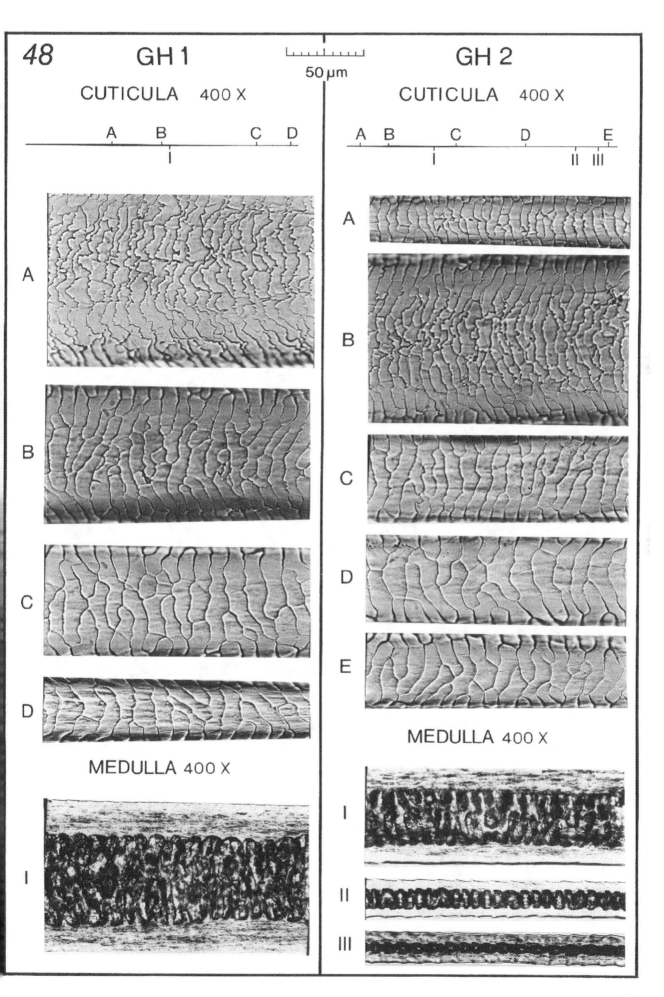

48 GH 1

CUTICULA 400 X

A   B   C D
I

A

B

C

D

MEDULLA 400 X

I

GH 2

CUTICULA 400 X

A B   C   D   E
II III

A

B

C

D

E

MEDULLA 400 X

I

II

III

50 µm

49 *Sciurus vulgaris*

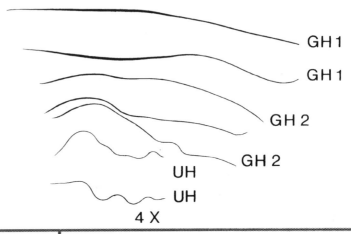

GH 1
GH 1
GH 2
GH 2
UH
UH

4 X

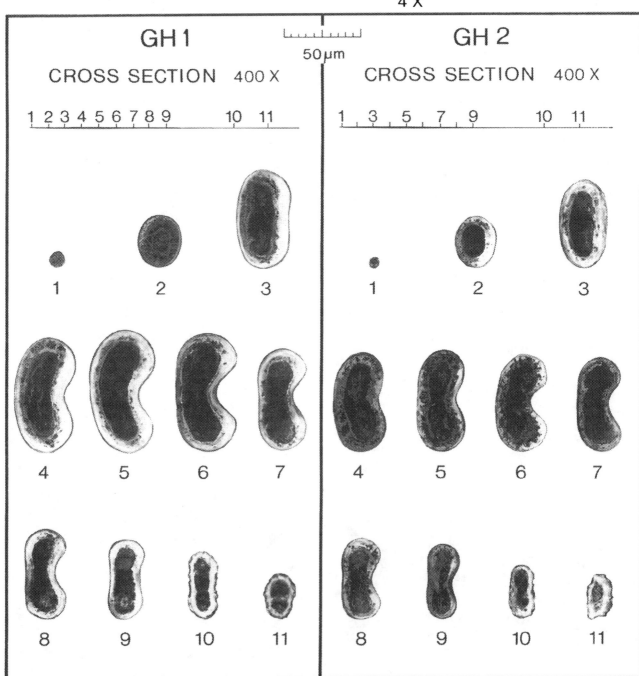

GH 1

CROSS SECTION 400 X

1 2 3 4 5 6 7 8 9   10   11

GH 2

CROSS SECTION 400 X

1 3 5 7 9   10   11

50 µm

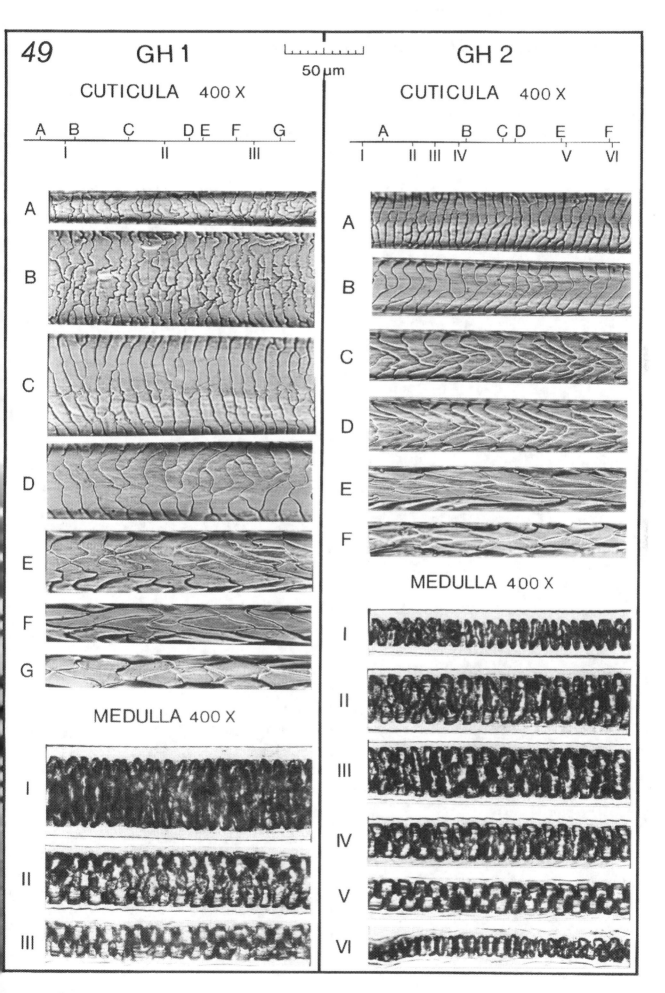

49   GH 1   50 μm   GH 2

CUTICULA 400 X   CUTICULA 400 X

MEDULLA 400 X

50 *Sciurus carolinensis*

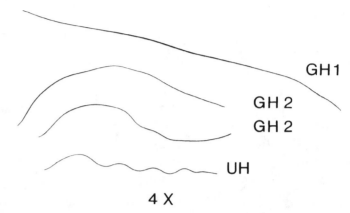

GH1

GH 2

GH 2

UH

4 X

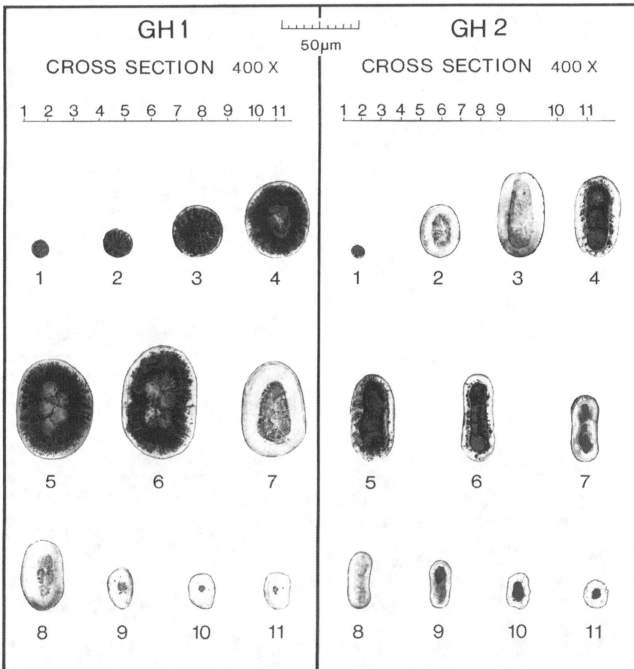

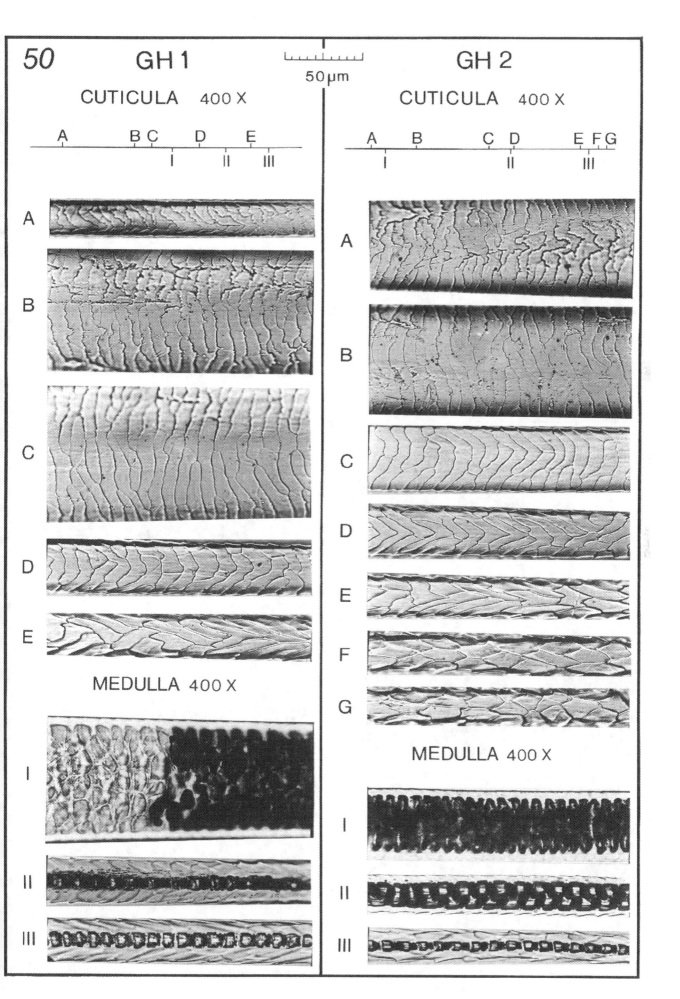

50 GH 1

CUTICULA 400 X

A B C D E
I II III

A
B
C
D
E

MEDULLA 400 X

I
II
III

GH 2

CUTICULA 400 X

A B C D E F G
I II III

A
B
C
D
E
F
G

MEDULLA 400 X

I
II
III

51 Tamias sibiricus

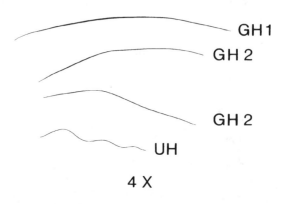

GH 1

GH 2

GH 2

UH

4 X

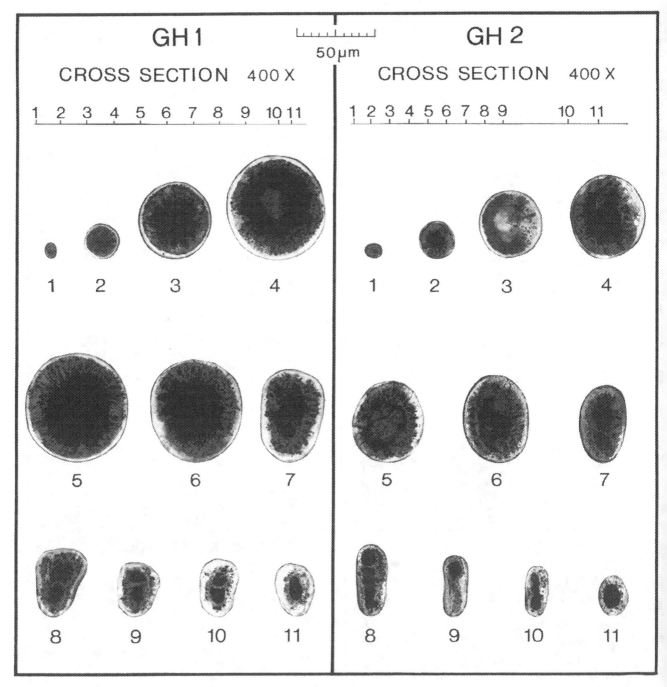

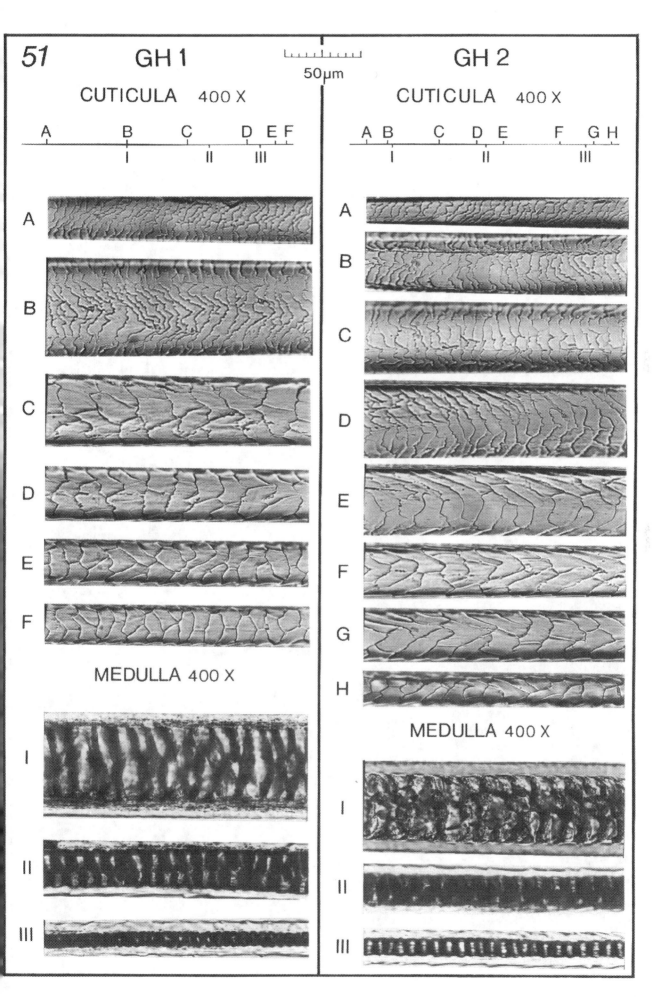

51 GH 1

CUTICULA 400 X

GH 2

CUTICULA 400 X

MEDULLA 400 X

MEDULLA 400 X

50µm

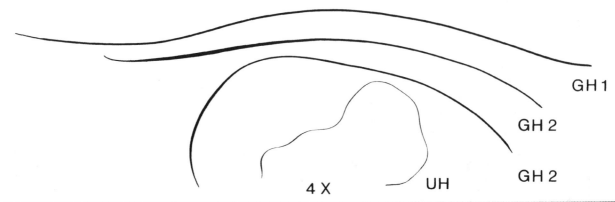

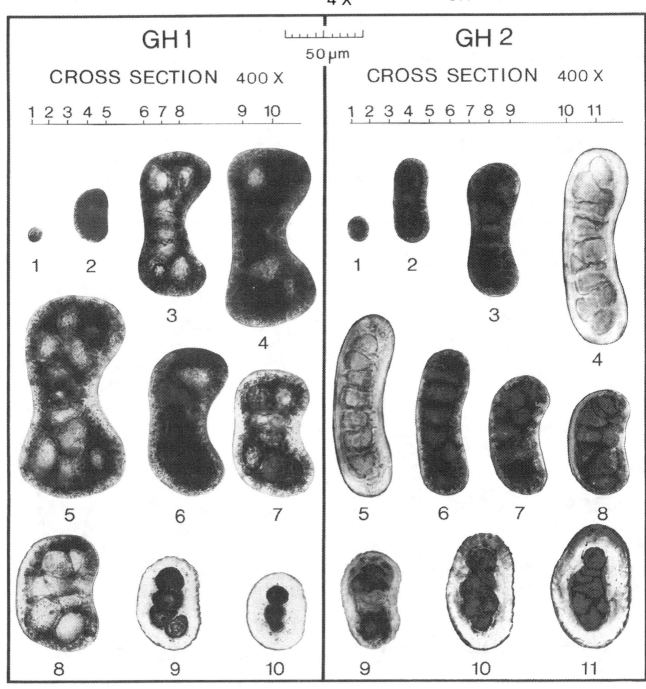

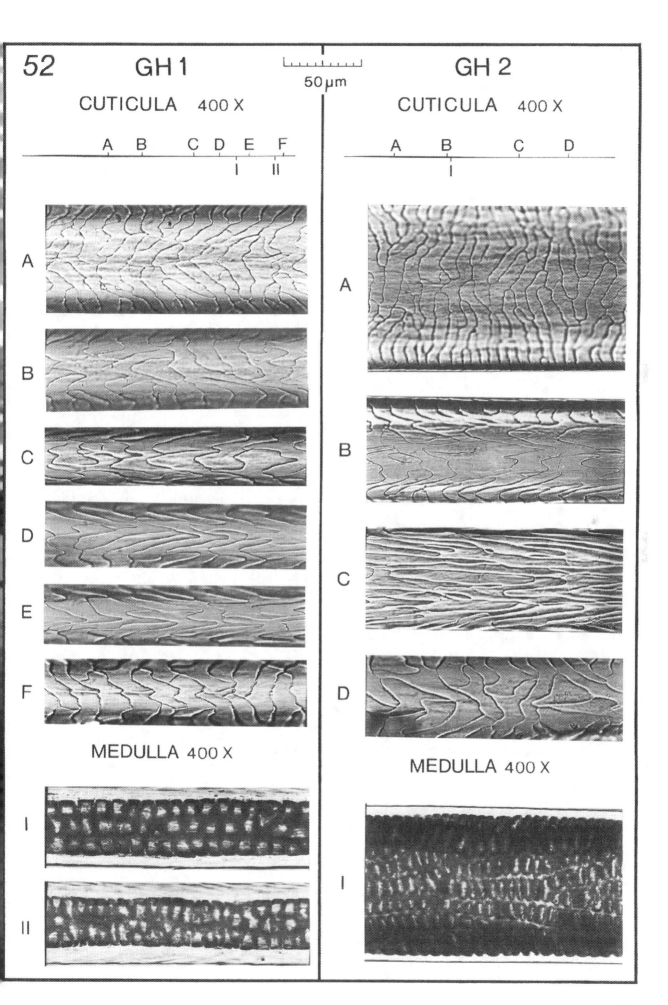

52　　GH 1

CUTICULA　400 X

50 µm

A　B　　C D E F
　　　　　　　I　II

A

B

C

D

E

F

MEDULLA 400 X

I

II

GH 2

CUTICULA　400 X

A　　B　　C　　D
　　　I

A

B

C

D

MEDULLA 400 X

I

53 Lepus timidus

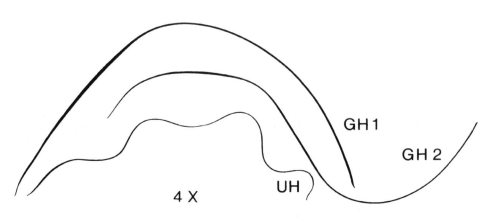

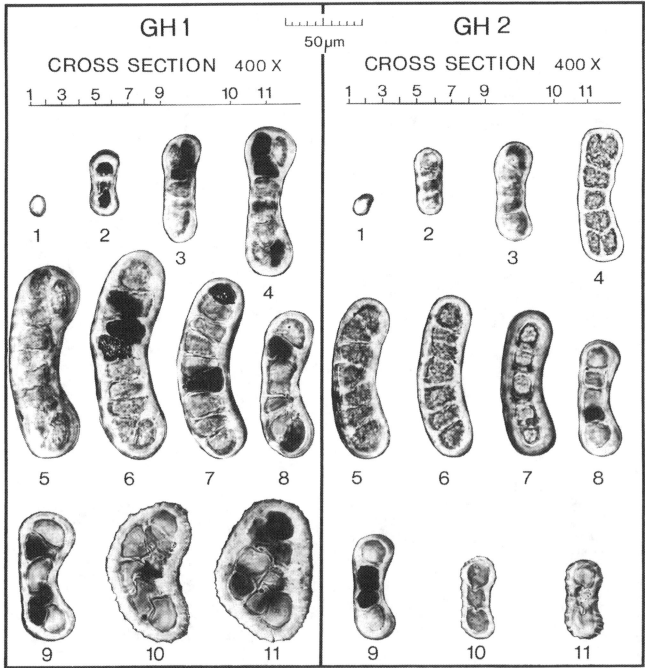

GH 1 CROSS SECTION 400 X

GH 2 CROSS SECTION 400 X

50 µm

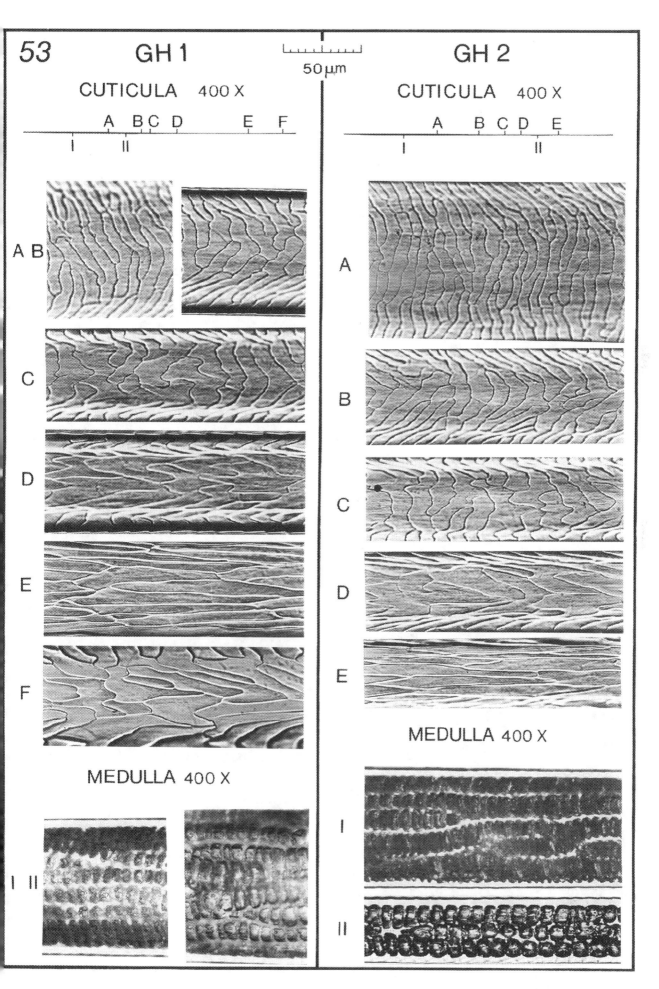

53 GH 1

CUTICULA 400 X

A B C D   E   F
I   II

A B

C

D

E

F

MEDULLA 400 X

I II

GH 2

CUTICULA 400 X

A   B C D E
I   II

A

B

C

D

E

MEDULLA 400 X

I

II

177

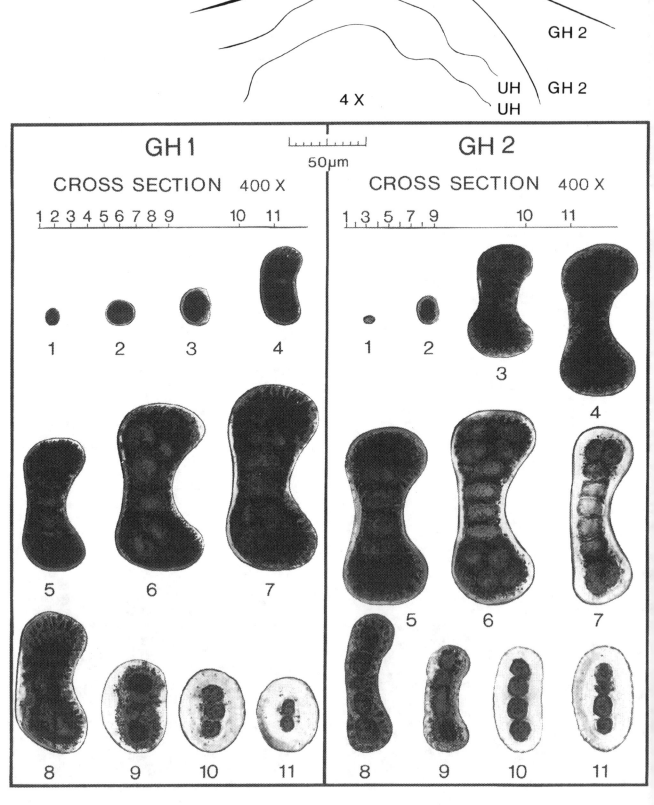

GH1

GH2

UH GH2

UH

4 X

GH1

50µm

CROSS SECTION 400 X

1 2 3 4 5 6 7 8 9 10 11

1 2 3 4 5 6 7 8 9 10 11

GH2

CROSS SECTION 400 X

1 3 5 7 9 10 11

1 2 3 4 5 6 7 8 9 10 11

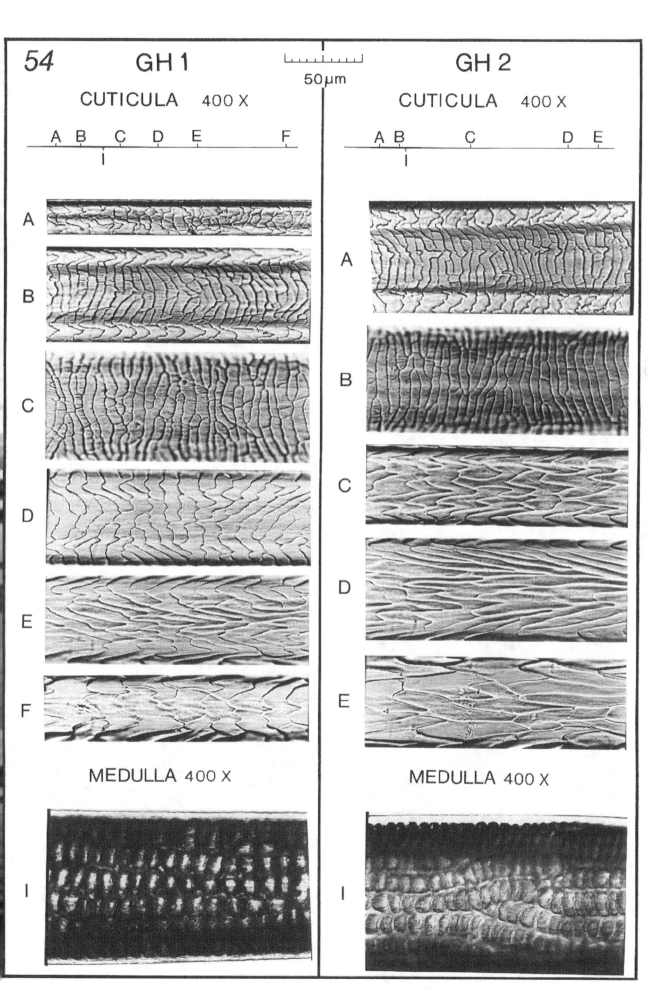

54 GH 1

CUTICULA 400 X

A B C D E F

GH 2

CUTICULA 400 X

A B C D E

50 µm

A

B

C

D

E

F

A

B

C

D

E

MEDULLA 400 X

MEDULLA 400 X

I

I

55 *Vulpes vulpes*

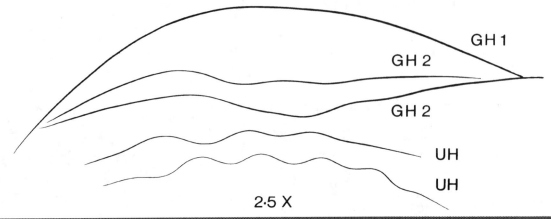

GH 1
GH 2
GH 2
UH
UH

2·5 X

GH 1	GH 2

100 μm

CROSS SECTION 200 X

CROSS SECTION 200 X

1 2 3 4 5 6 7 8 9   10   11

1 3 5 7 9   10   11

1

2

3

4

5

6

7

8

9

10

11

1

2

3

4

5

6

7

8

9

10

11

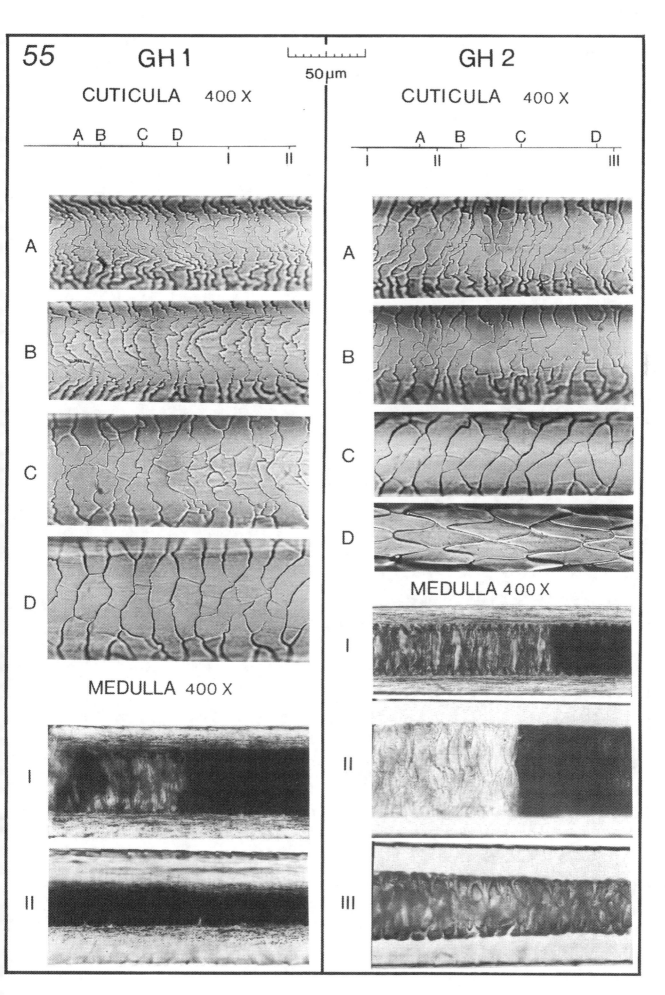

55   GH 1

CUTICULA   400 X

A   B   C   D

I   II

50 μm

A

B

C

D

MEDULLA   400 X

I

II

GH 2

CUTICULA   400 X

A   B   C   D

I   II   III

A

B

C

D

MEDULLA 400 X

I

II

III

56 Canis familiaris

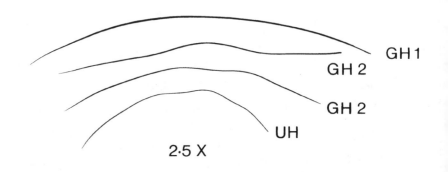

GH 1
GH 2
GH 2
UH
2·5 X

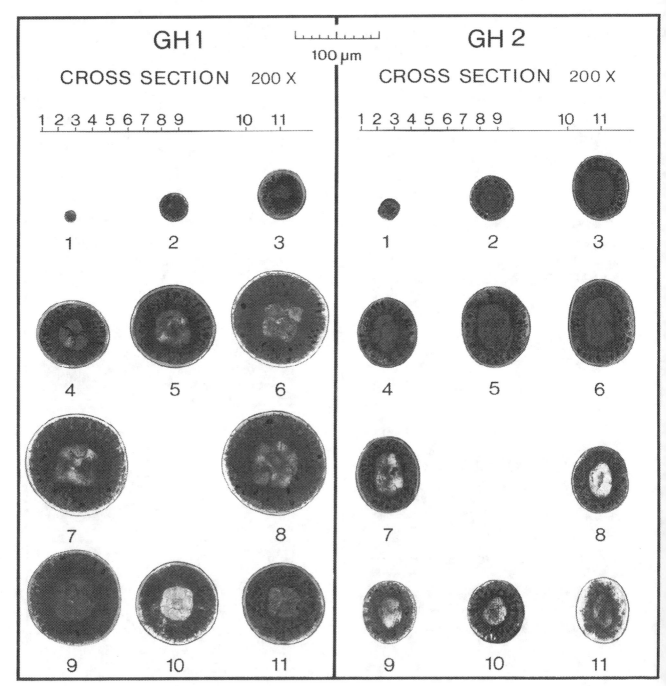

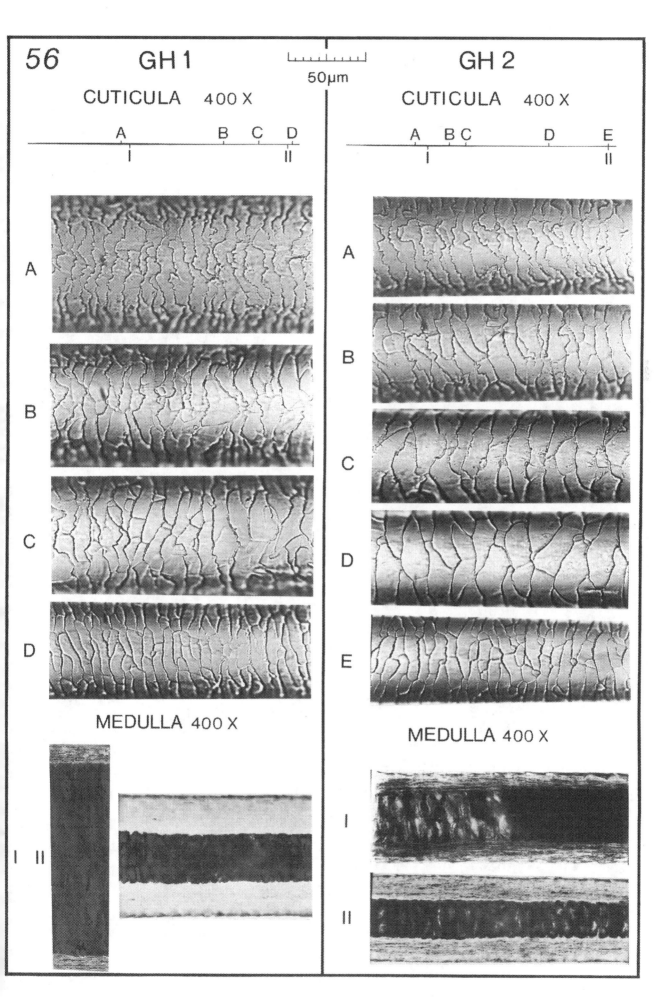

56 GH 1

CUTICULA 400 X

A B C D
I II

A

B

C

D

MEDULLA 400 X

I II

GH 2

CUTICULA 400 X

A B C D E
I II

A

B

C

D

E

MEDULLA 400 X

I

II

57 Nyctereutes procyonoides

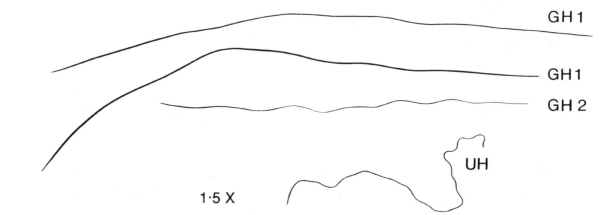

GH 1

GH 1

GH 2

UH

1·5 X

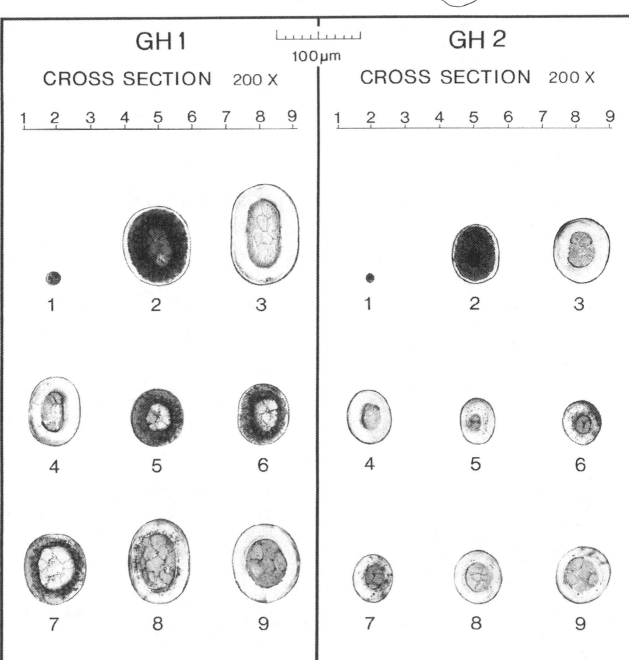

GH 1 | 100 µm | GH 2

CROSS SECTION 200 X

CROSS SECTION 200 X

1 2 3 4 5 6 7 8 9

1 2 3 4 5 6 7 8 9

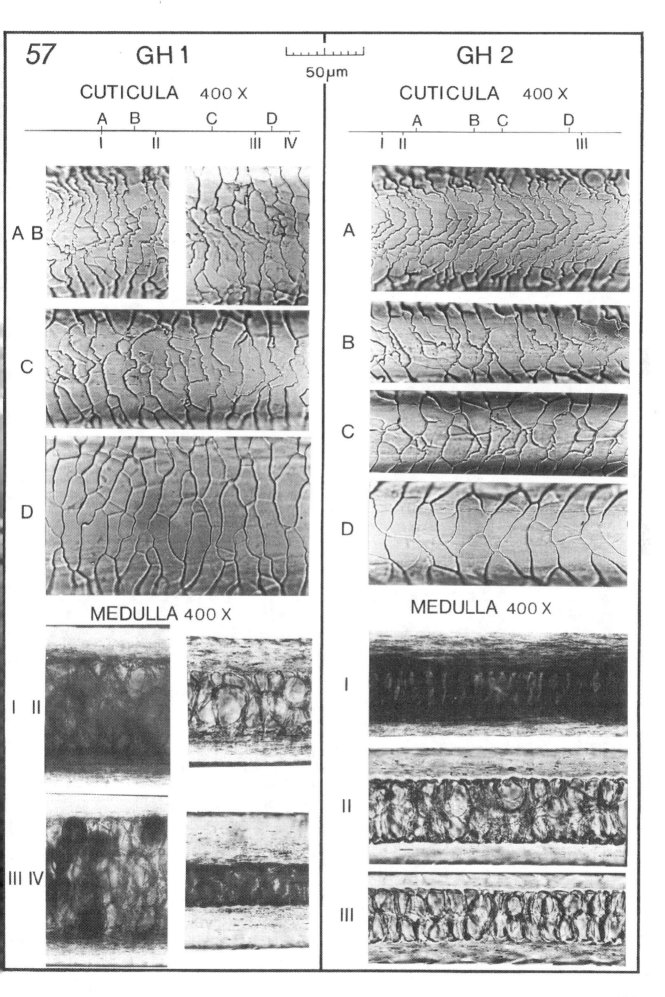

57 GH 1

CUTICULA 400 X

A B   C   D
  I   II   III   IV

A B

C

D

MEDULLA 400 X

I II

III IV

GH 2

CUTICULA 400 X

A   B C   D
I II   III

A

B

C

D

MEDULLA 400 X

I

II

III

185

58 Procyon lotor

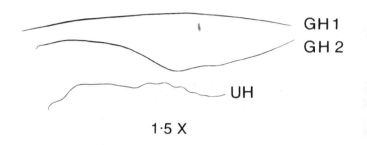

GH1
GH2
UH

1·5 X

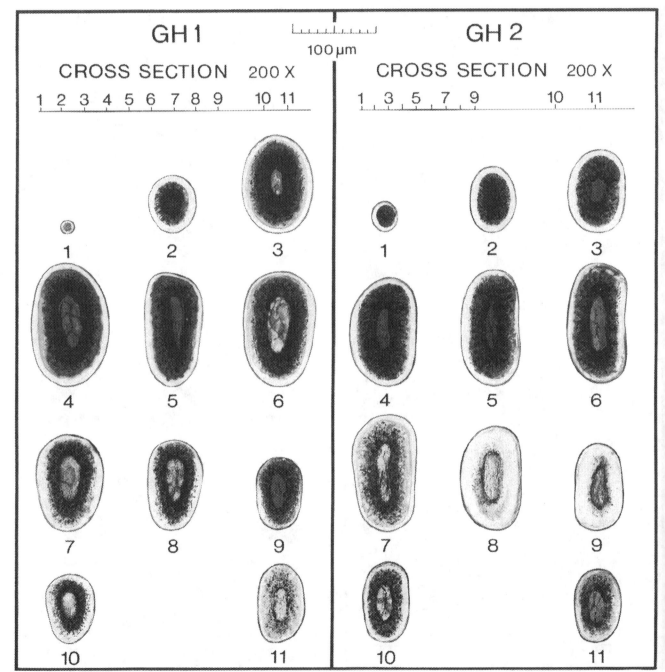

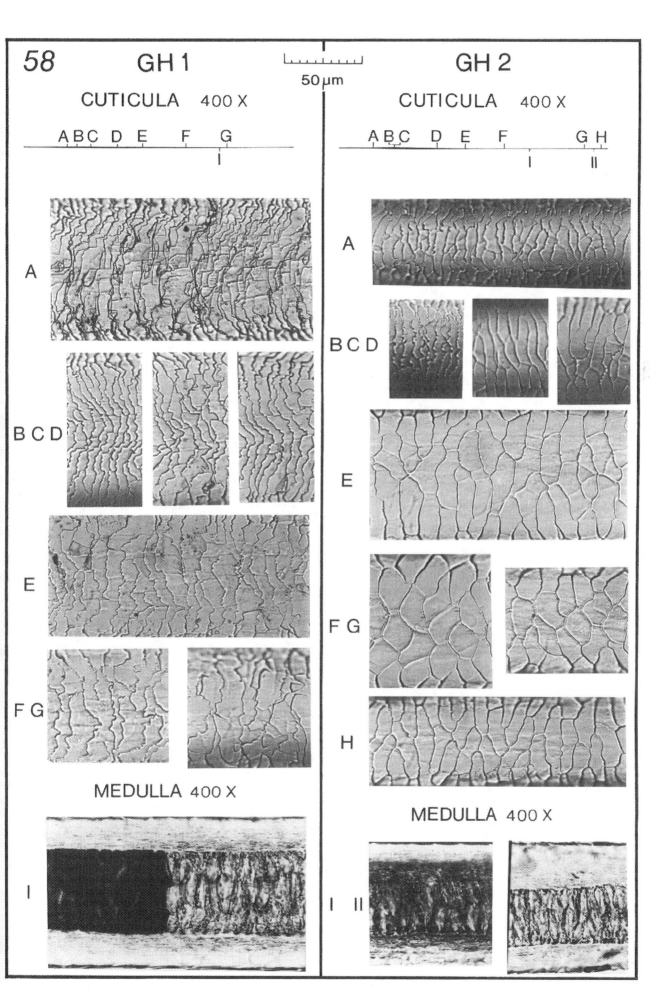

58   GH 1

CUTICULA 400 X

A B C D E F G

GH 2

CUTICULA 400 X

A B C D E F G H

A

B C D

E

F G

MEDULLA 400 X

I

A

B C D

E

F G

H

MEDULLA 400 X

I II

59 Mustela erminea

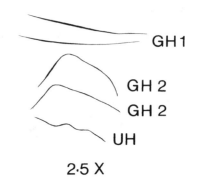

GH1
GH 2
GH 2
UH

2·5 X

GH 1

100 µm

CROSS SECTION   200 X

1 2 3 4 5 6 7 8 9 10 11

1   2   3

4   5   6

7   8   9

10   11

GH 2

CROSS SECTION   200 X

1 2 3 4 5 6 7 8 9   10   11

1   2   3

4   5   6

7   8   9

10   11

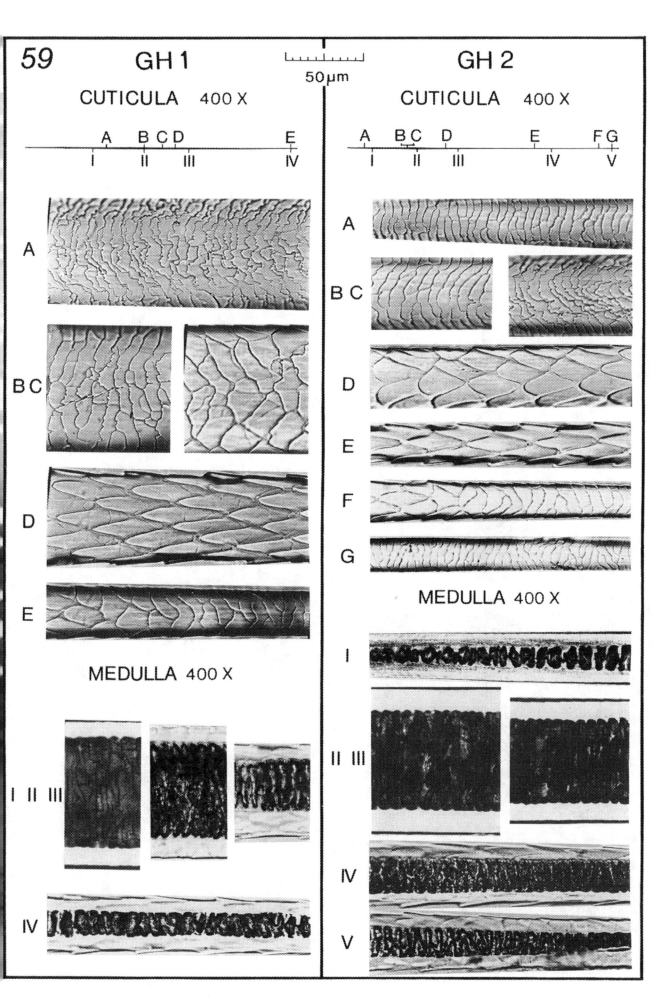

59   GH 1

CUTICULA   400 X

A   B C D   E
 I   II   III   IV

A

B C

D

E

MEDULLA 400 X

I II III

IV

GH 2

CUTICULA   400 X

A   B C   D   E   F G
 I   II   III   IV   V

A

B C

D

E

F

G

MEDULLA 400 X

I

II III

IV

V

60 Mustela nivalis

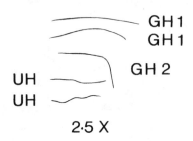

GH 1
GH 1
GH 2
UH
UH

2·5 X

GH 1	GH 2
CROSS SECTION 200 X	CROSS SECTION 200 X

100 µm

1 2 3 4 5 6 7 8 9 10 11

1 2 3 4 5 6 7 8 9 10 11

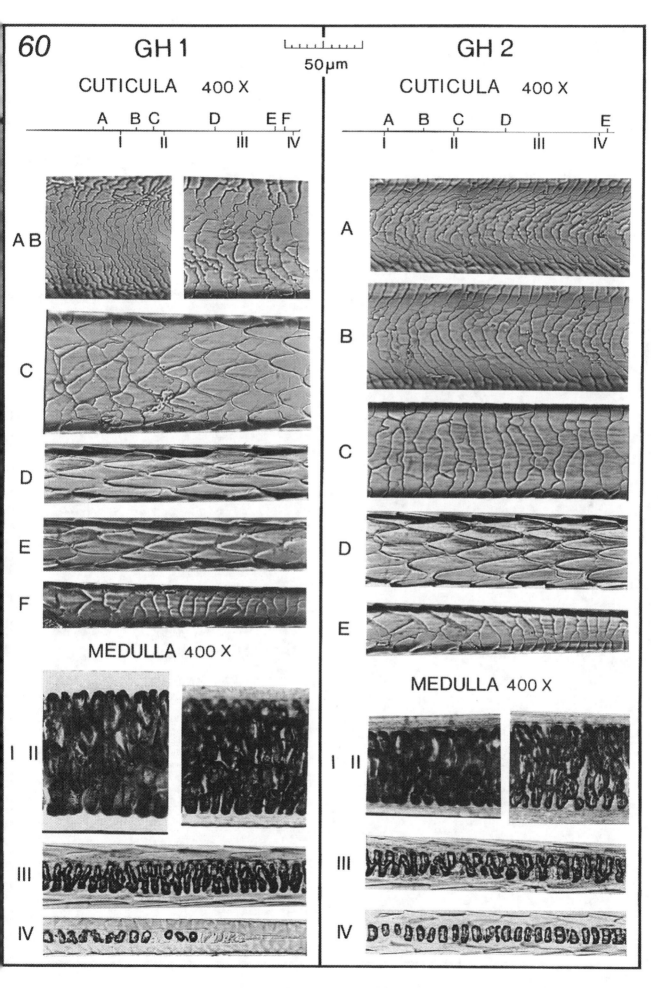

60

GH 1

CUTICULA 400 X

A B C   D   E F

I   II   III   IV

50 µm

A B

C

D

E

F

MEDULLA 400 X

I II

III

IV

GH 2

CUTICULA 400 X

A   B   C   D   E

I   II   III   IV

A

B

C

D

E

MEDULLA 400 X

I II

III

IV

61 *Mustela putorius*

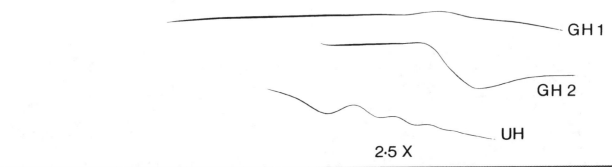

GH 1

GH 2

UH

2·5 X

GH 1

CROSS SECTION 200 X

100µm

GH 2

CROSS SECTION 200 X

1 2 3 4 5 6 7 8 9   10   11

1 3 5 7 9   10   11

1
2
3
4
5
6
7
8
9
10
11

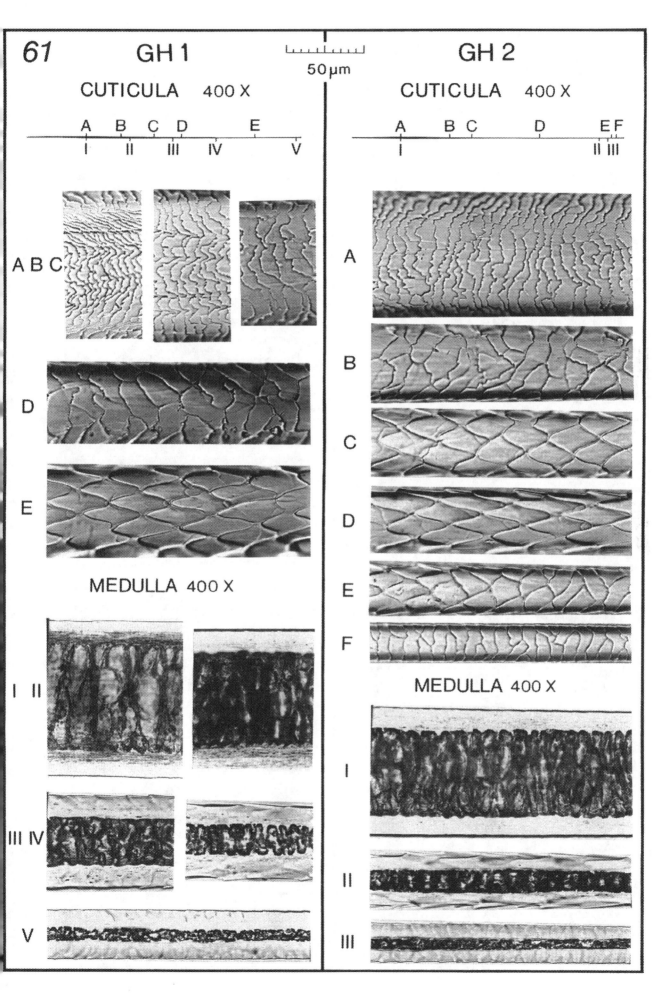

61 GH 1

CUTICULA 400 X

GH 2

CUTICULA 400 X

50 µm

MEDULLA 400 X

MEDULLA 400 X

193

62 *Mustela vison*

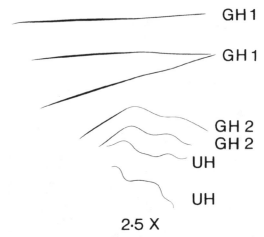

GH 1
GH 1
GH 2
GH 2
UH
UH

2·5 X

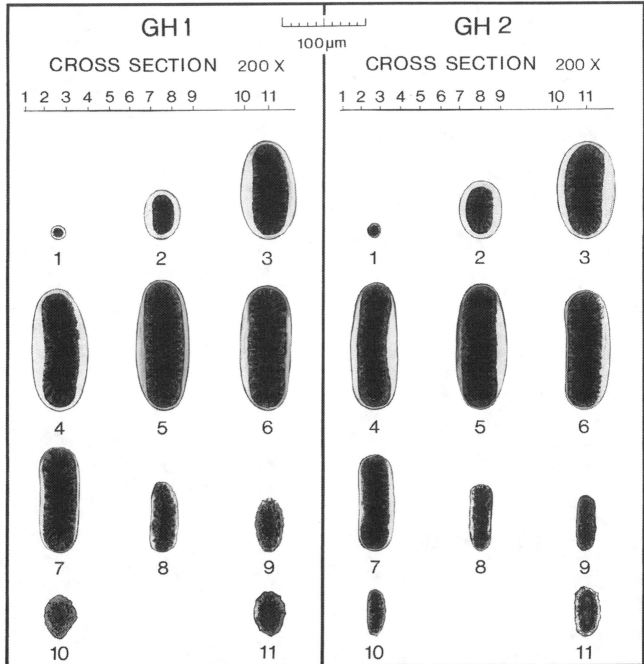

GH 1
CROSS SECTION 200 X
1 2 3 4 5 6 7 8 9 10 11

100 µm

GH 2
CROSS SECTION 200 X
1 2 3 4 5 6 7 8 9 10 11

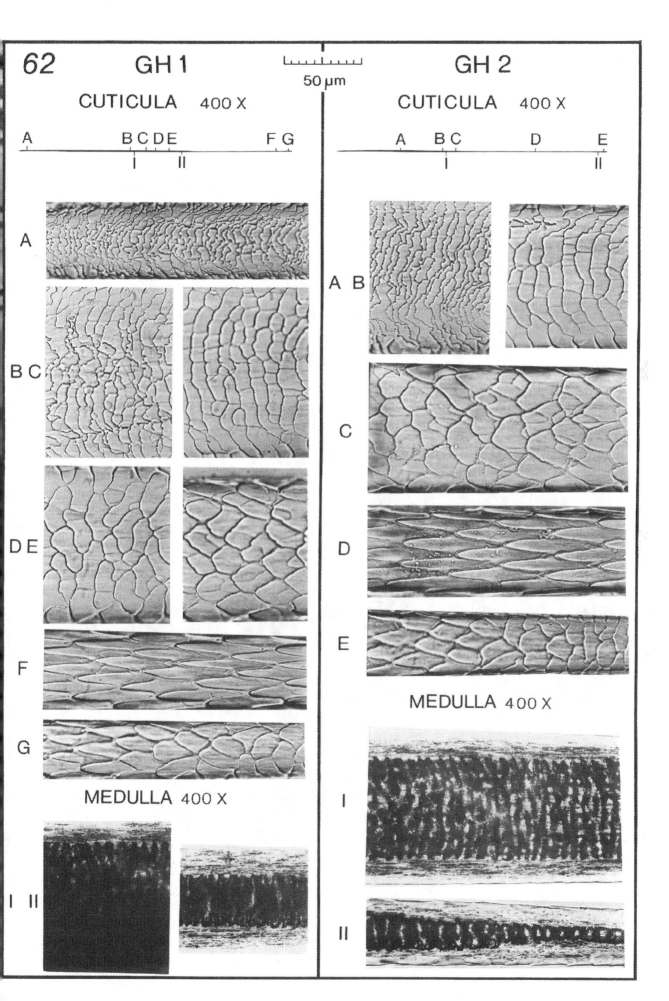

62 GH 1 GH 2

50 µm

CUTICULA 400 X

CUTICULA 400 X

195

63 Martes martes

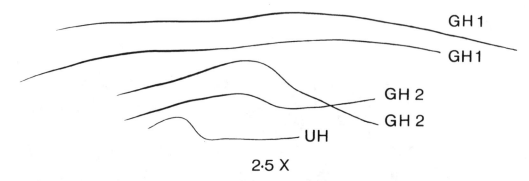

GH 1
GH 1
GH 2
GH 2
UH

2·5 X

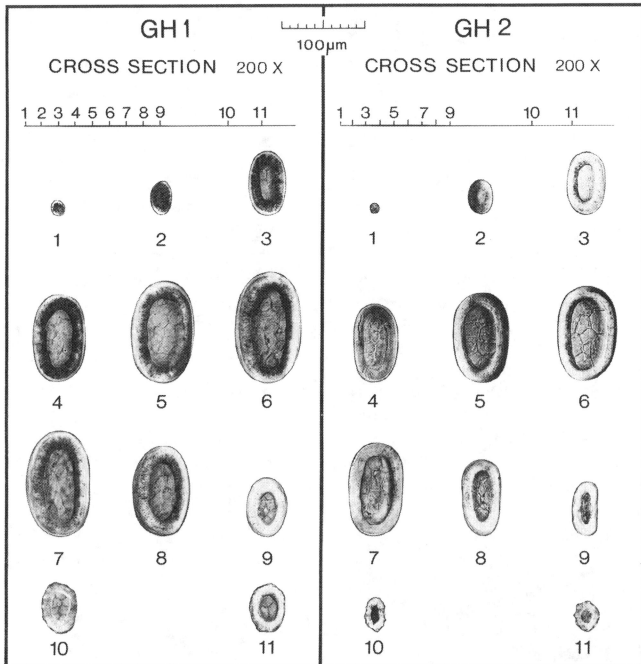

GH 1	GH 2
CROSS SECTION 200 X	CROSS SECTION 200 X

100µm

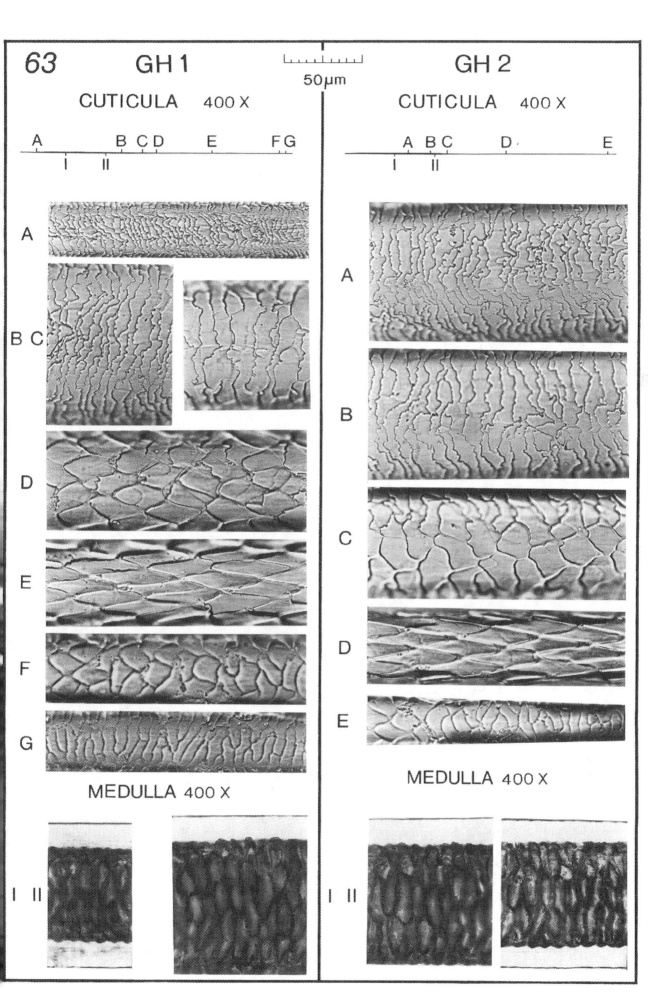

63   GH 1

CUTICULA   400 X

50µm

A   B C D   E   F G
I   II

A

B C

D

E

F

G

MEDULLA 400 X

I II

GH 2

CUTICULA   400 X

A B C   D   E
I   II

A

B

C

D

E

MEDULLA 400 X

I II

64 Martes foina

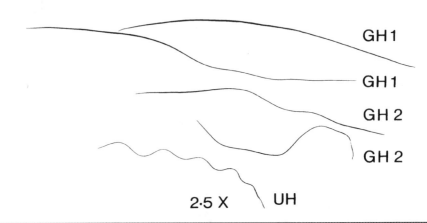

GH 1

GH 1

GH 2

GH 2

2·5 X   UH

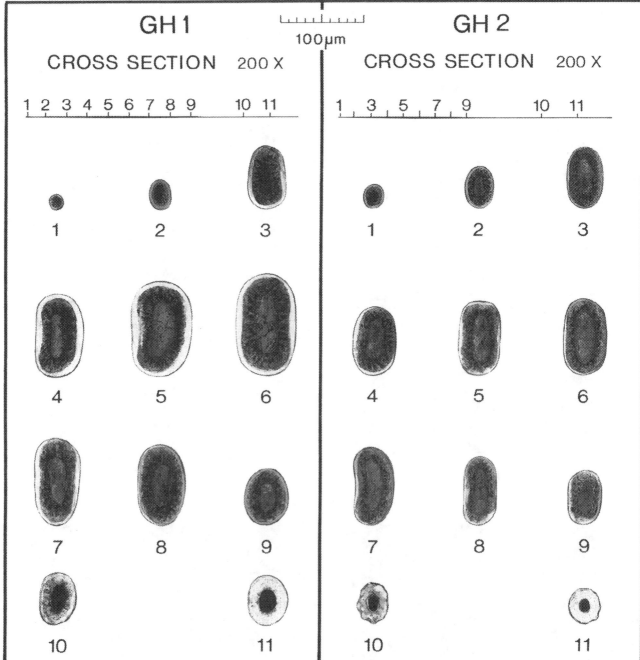

GH 1

GH 2

100 μm

CROSS SECTION   200 X

CROSS SECTION   200 X

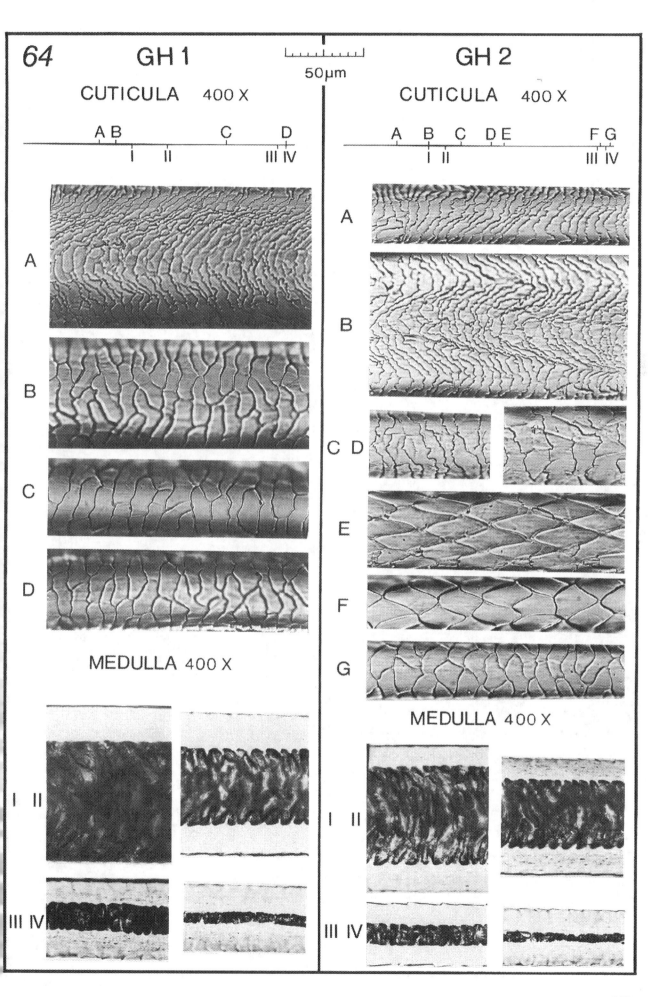

64   GH 1

CUTICULA   400 X

A B   C   D
    I   II   III IV

50µm

GH 2

CUTICULA   400 X

A   B   C   DE   F G
     I   II   III IV

A

B

C

D

MEDULLA 400 X

I II

III IV

A

B

C D

E

F

G

MEDULLA 400 X

I II

III IV

65 *Meles meles*

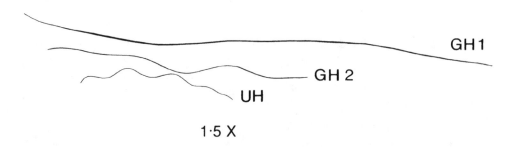

GH 1

GH 2

UH

1·5 X

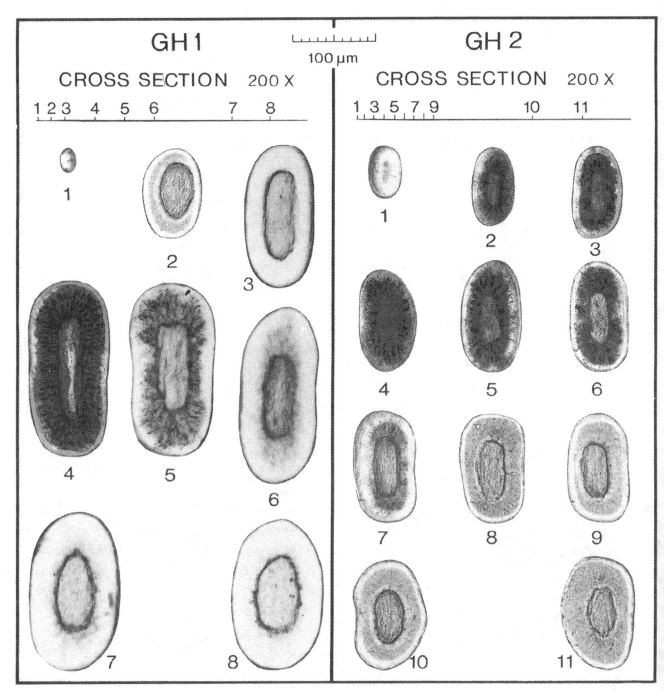

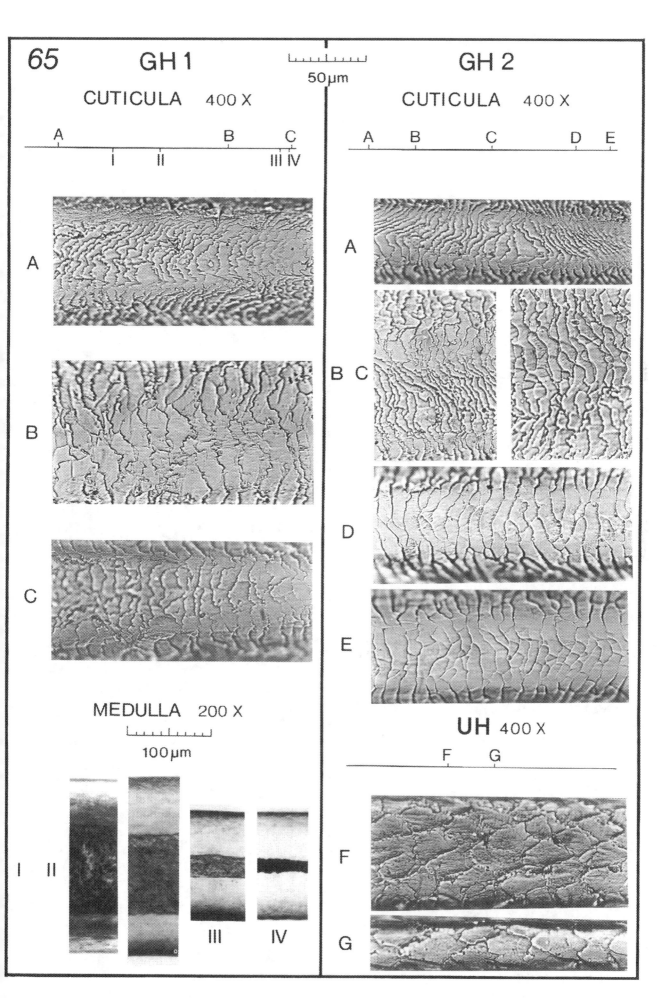

65　　GH 1

CUTICULA　400 X

50 µm

A　　　　　　B　　C
　　　I　　II　　　III IV

A

B

C

MEDULLA　200 X

100 µm

I　II　　　III　　IV

GH 2

CUTICULA　400 X

A　　B　　C　　　D　E

A

B C

D

E

UH　400 X

F　G

F

G

66 Lutra lutra

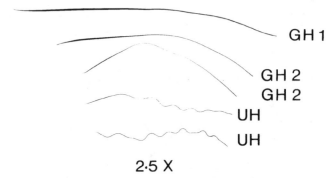

GH 1

GH 2
GH 2
UH
UH

2·5 X

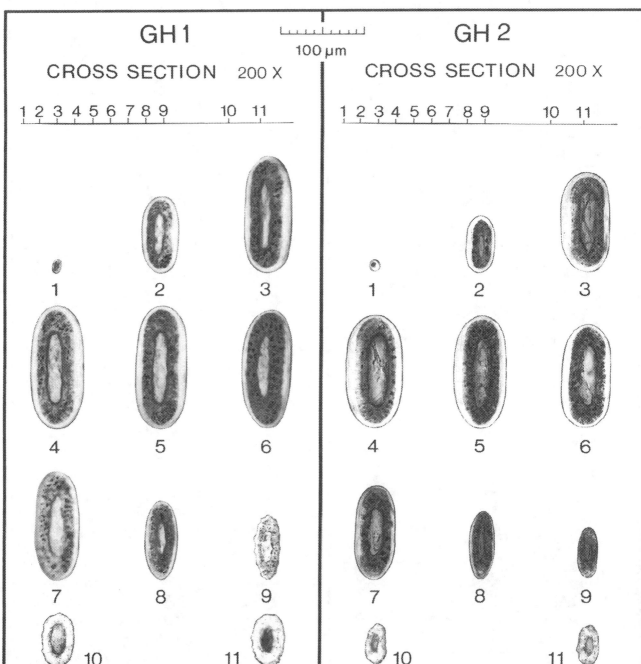

GH 1

CROSS SECTION 200 X

1 2 3 4 5 6 7 8 9   10 11

GH 2

CROSS SECTION 200 X

1 2 3 4 5 6 7 8 9   10 11

100 µm

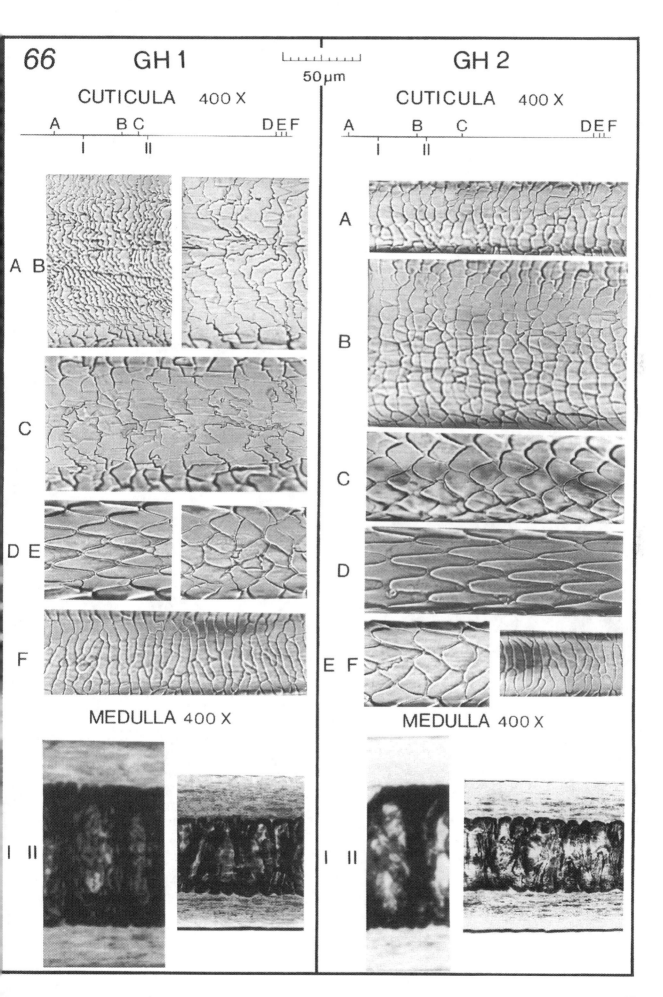

66   GH 1

CUTICULA   400 X

A   B C   D E F
      I   II

GH 2

CUTICULA   400 X

A   B   C   D E F
      I   II

A B

C

D E

F

A

B

C

D

E F

MEDULLA 400 X

I II

MEDULLA 400 X

I II

50 µm

67 Felis silvestris

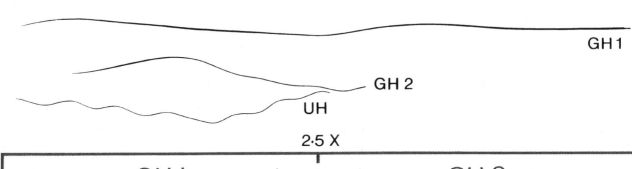

GH 1

GH 2

UH

2·5 X

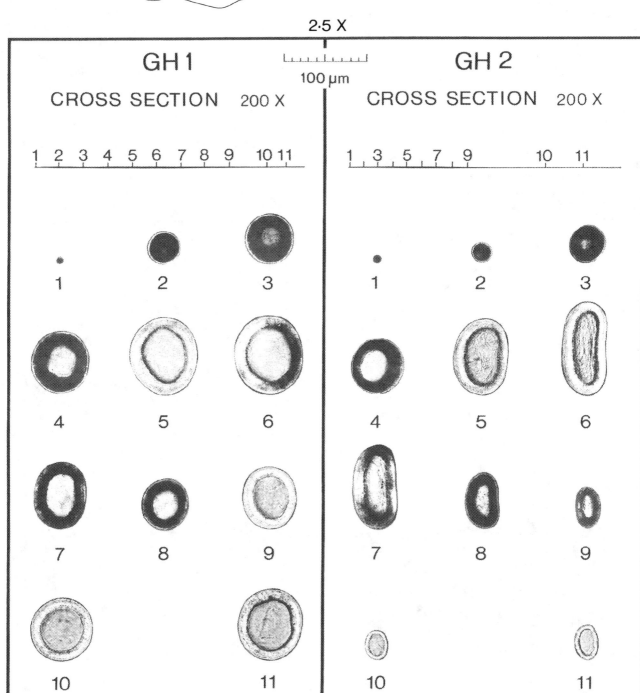

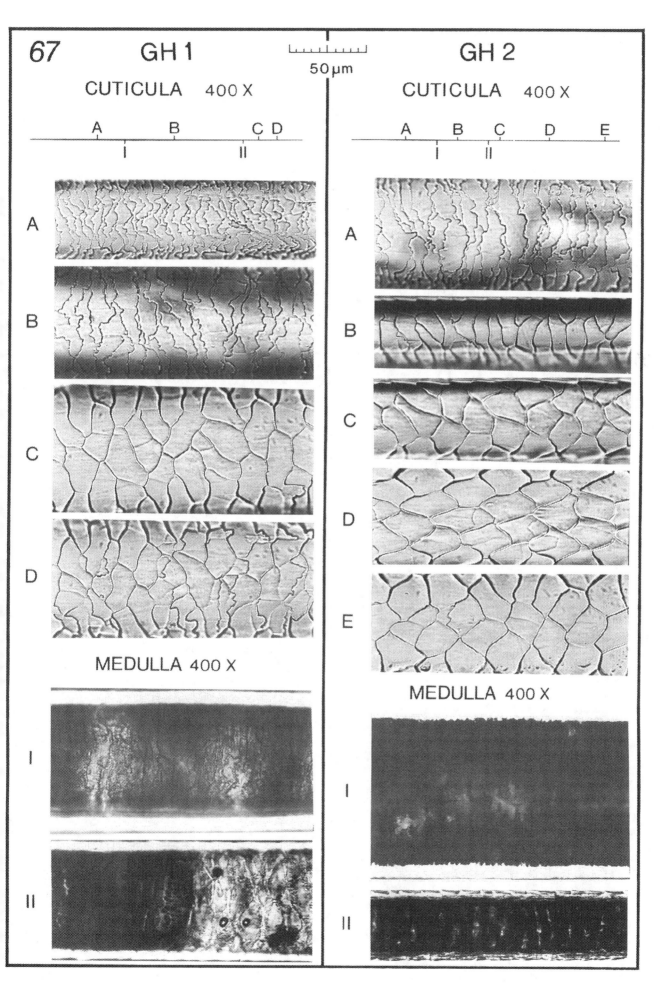

67 GH 1

CUTICULA 400 X

A   B   C D

I   II

A

B

C

D

MEDULLA 400 X

I

II

GH 2

CUTICULA 400 X

A   B   C   D   E

I   II

A

B

C

D

E

MEDULLA 400 X

I

II

50 µm

68 Felis catus

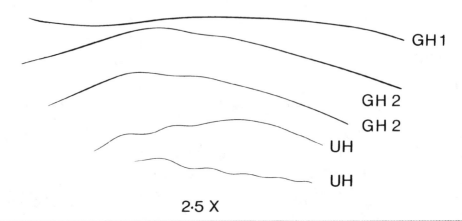

2·5 X

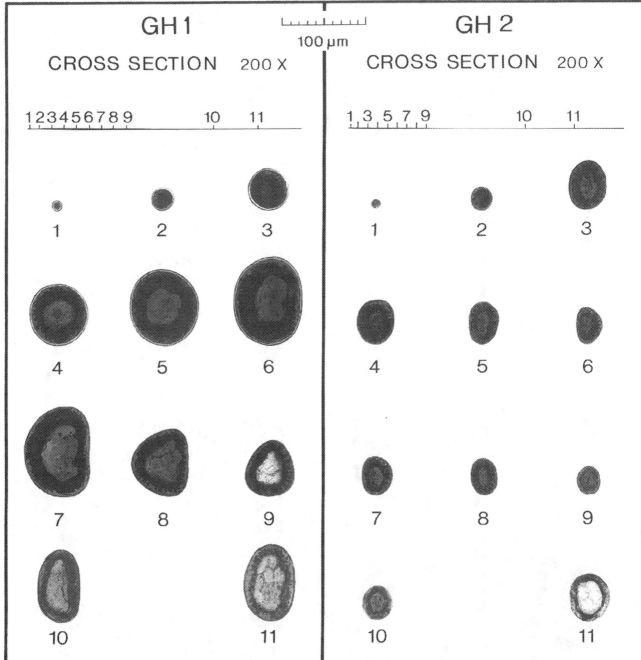

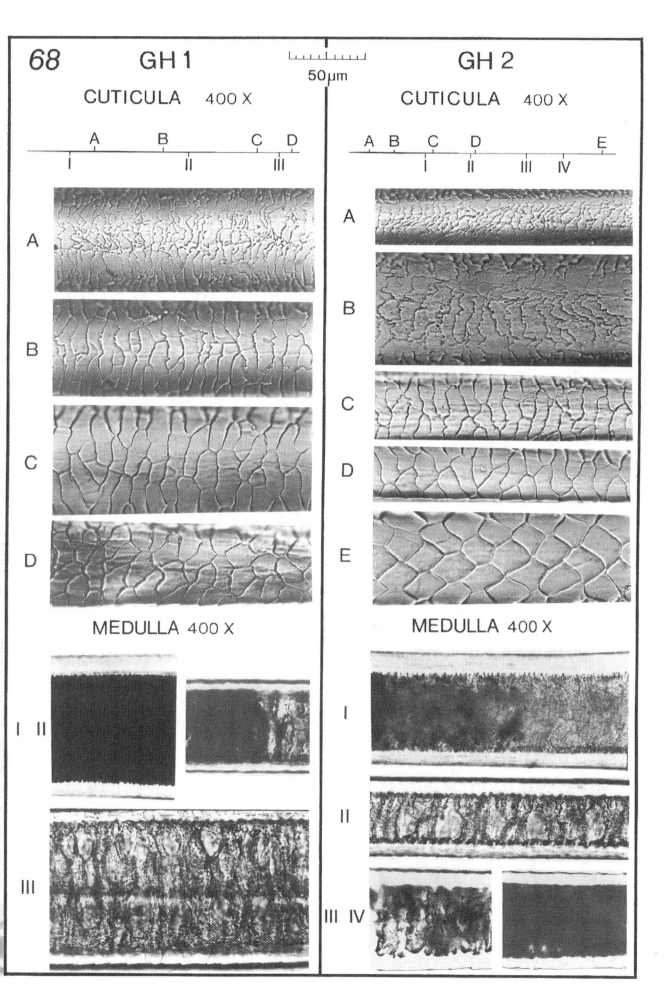

68 GH 1   GH 2

CUTICULA 400 X   CUTICULA 400 X

50 µm

MEDULLA 400 X   MEDULLA 400 X

69 Sus scrofa

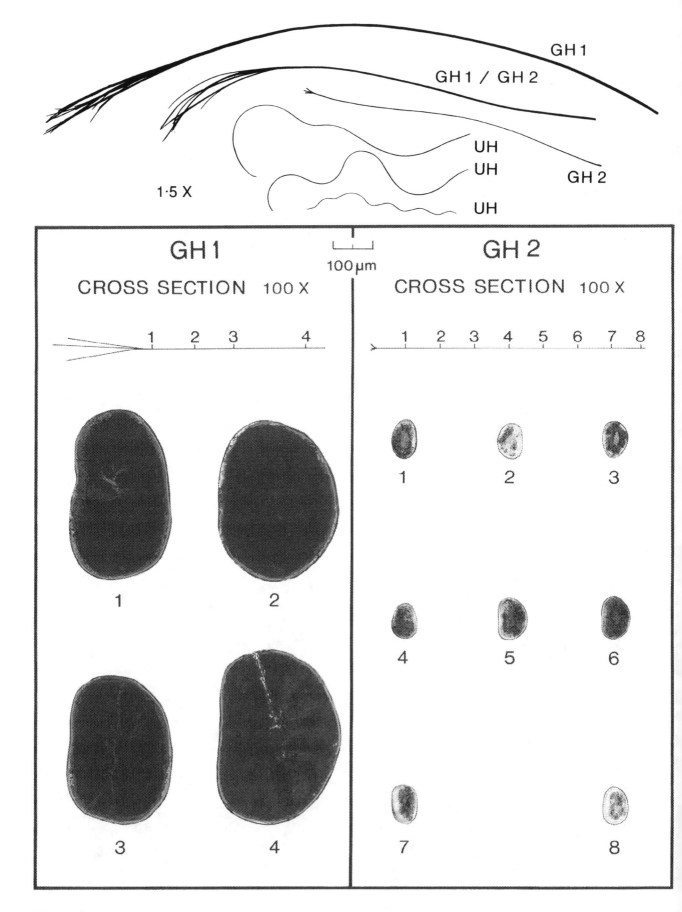

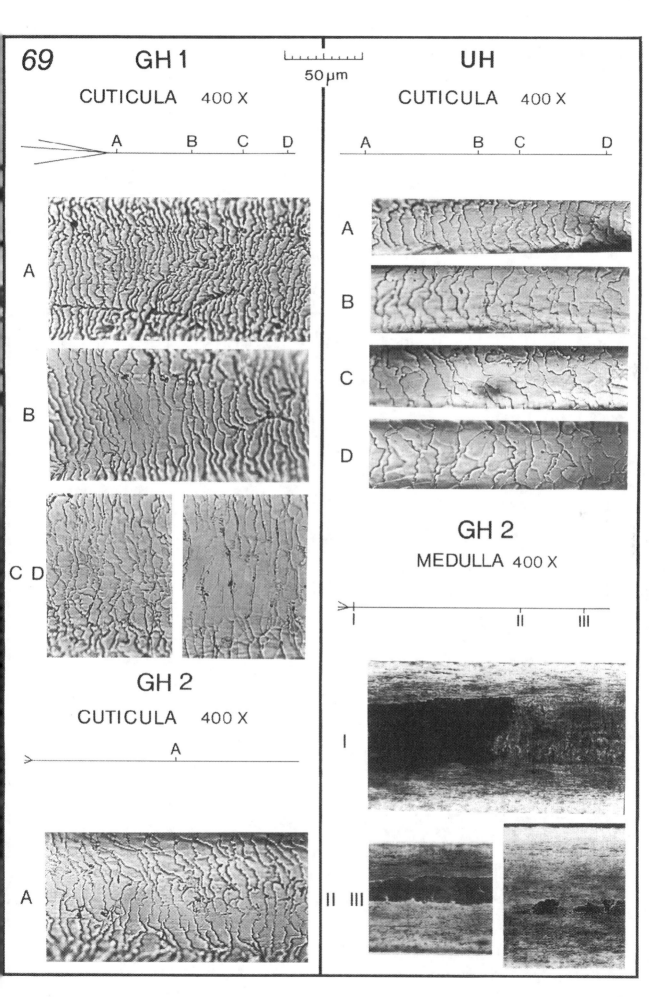

69 GH 1
CUTICULA 400 X
A B C D

A

B

C D

GH 2
CUTICULA 400 X
A

A

UH
CUTICULA 400 X
A B C D

A

B

C

D

GH 2
MEDULLA 400 X
I II III

I

II III

50 µm

70 Cervus dama

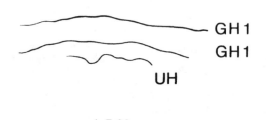

GH 1
GH 1
UH

1·5 X

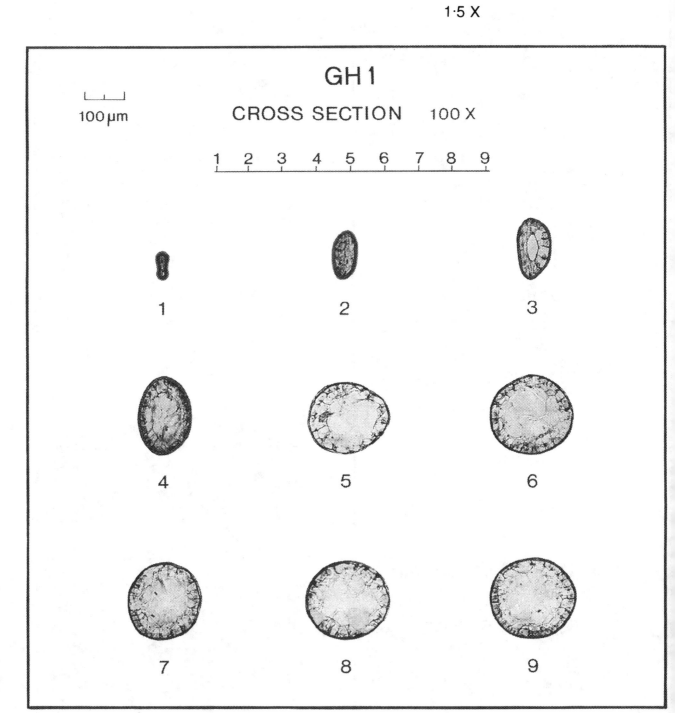

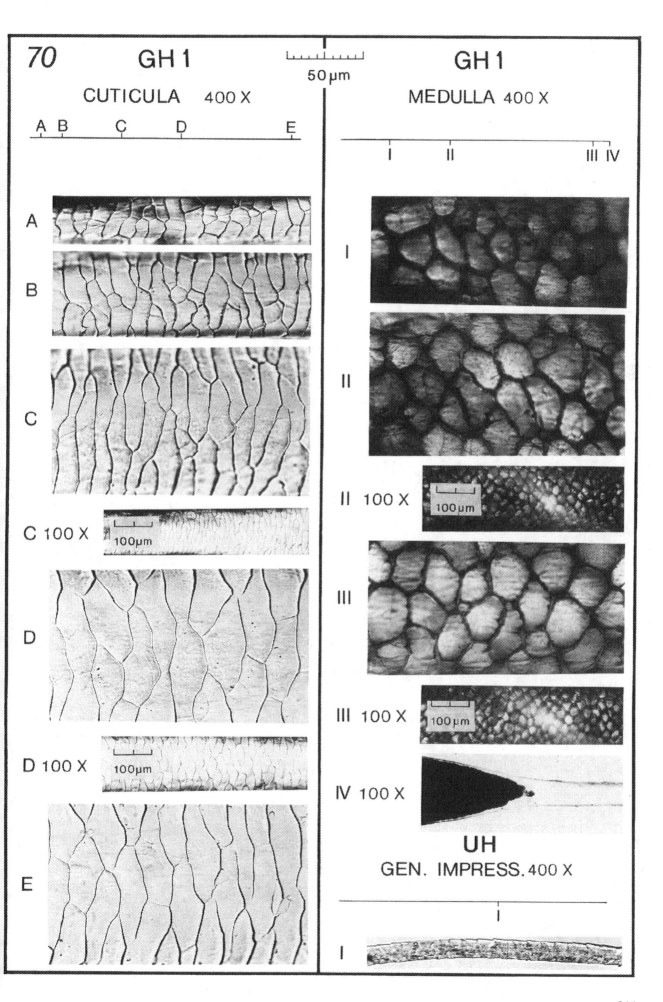

70 GH 1 CUTICULA 400 X 50 µm

GH 1 MEDULLA 400 X

UH GEN. IMPRESS. 400 X

71 Cervus   elaphus

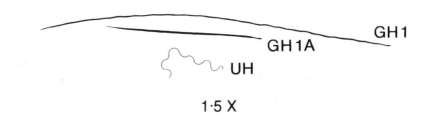

GH 1
GH 1A
UH

1·5 X

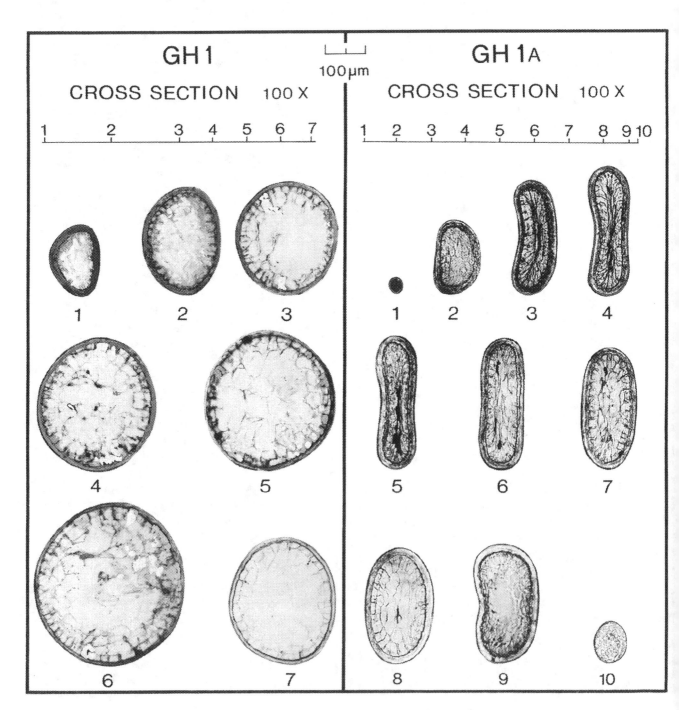

GH 1

CROSS SECTION   100 X

GH 1A

CROSS SECTION   100 X

100 µm

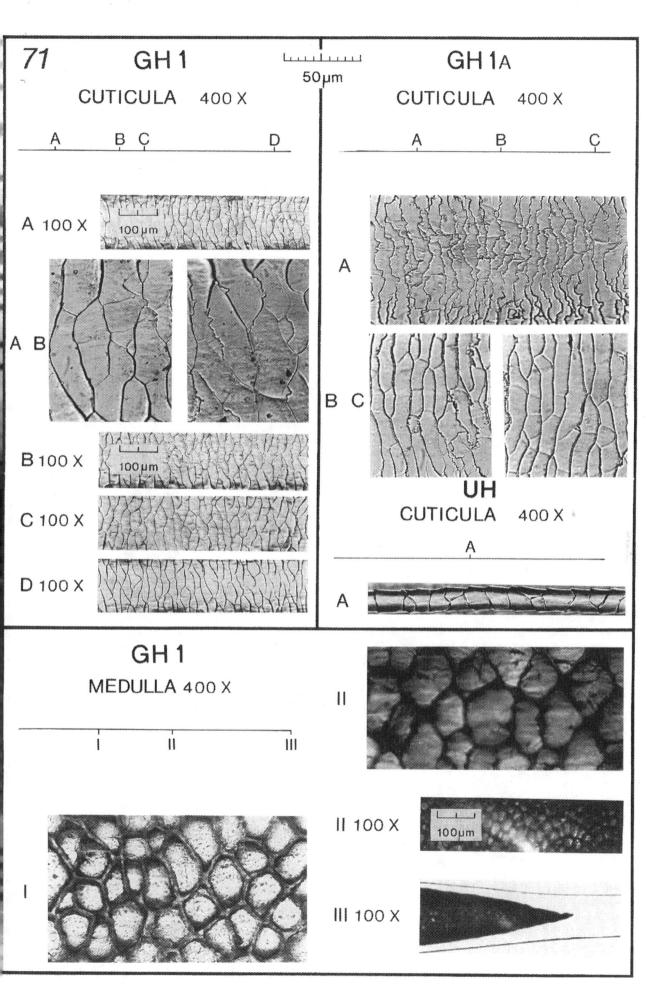

71 GH 1

CUTICULA 400 X

50 μm

A 100 X

A B

B 100 X

C 100 X

D 100 X

GH 1A

CUTICULA 400 X

B C

UH

CUTICULA 400 X

A

GH 1

MEDULLA 400 X

I II III

II

II 100 X

100 μm

III 100 X

I

213

72 *Capreolus capreolus*

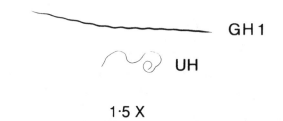

GH 1

UH

1·5 X

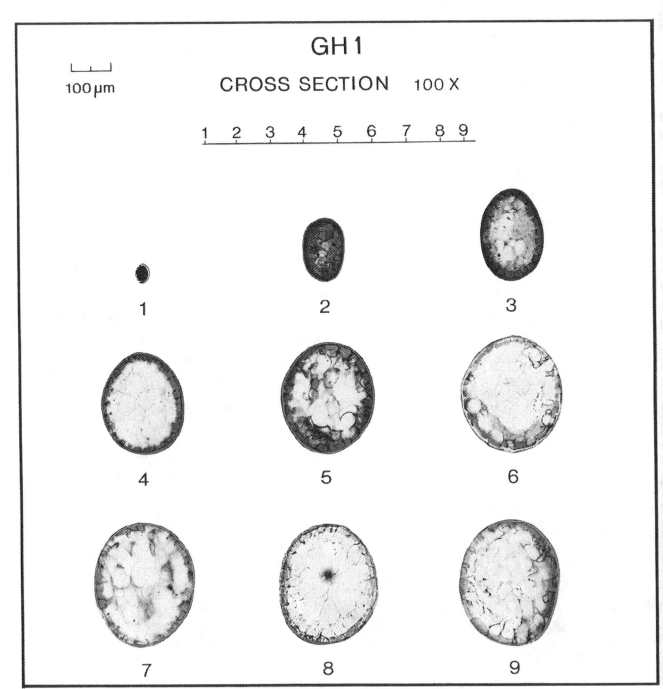

GH 1

CROSS SECTION 100 X

1 2 3 4 5 6 7 8 9

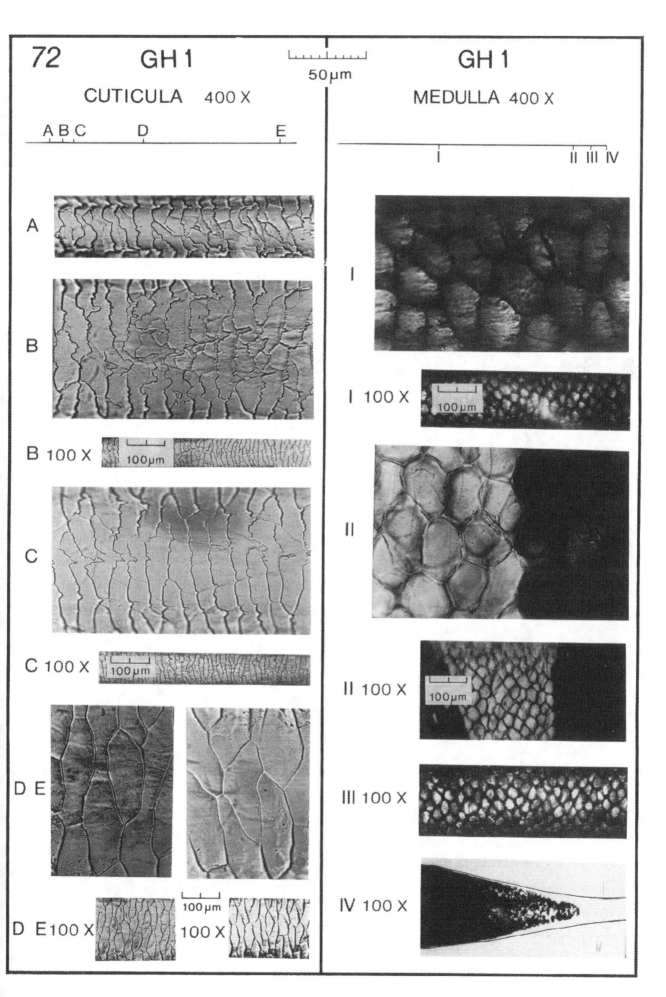

72 GH 1

CUTICULA 400 X

A B C　　　D　　　　　　　E

A

B

B 100 X　100µm

C

C 100 X　100µm

D E

D E 100 X　100 X　100µm

50µm

GH 1

MEDULLA 400 X

I　　　　II III IV

I

I 100 X　100 µm

II

II 100 X　100µm

III 100 X

IV 100 X

215

73 Ovis musimon

UH   GH1

1·5 X

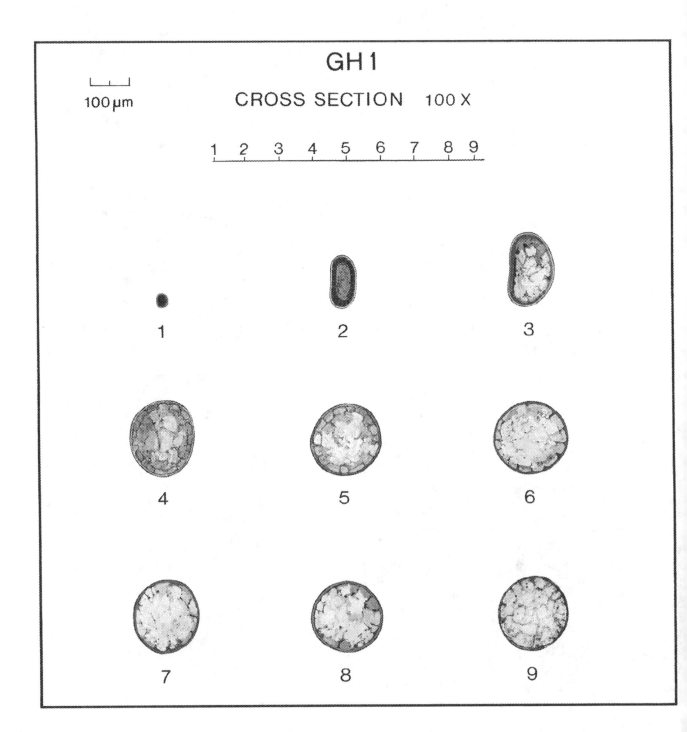

GH 1

CROSS SECTION   100 X

100 µm

1   2   3   4   5   6   7   8   9

1   2   3

4   5   6

7   8   9

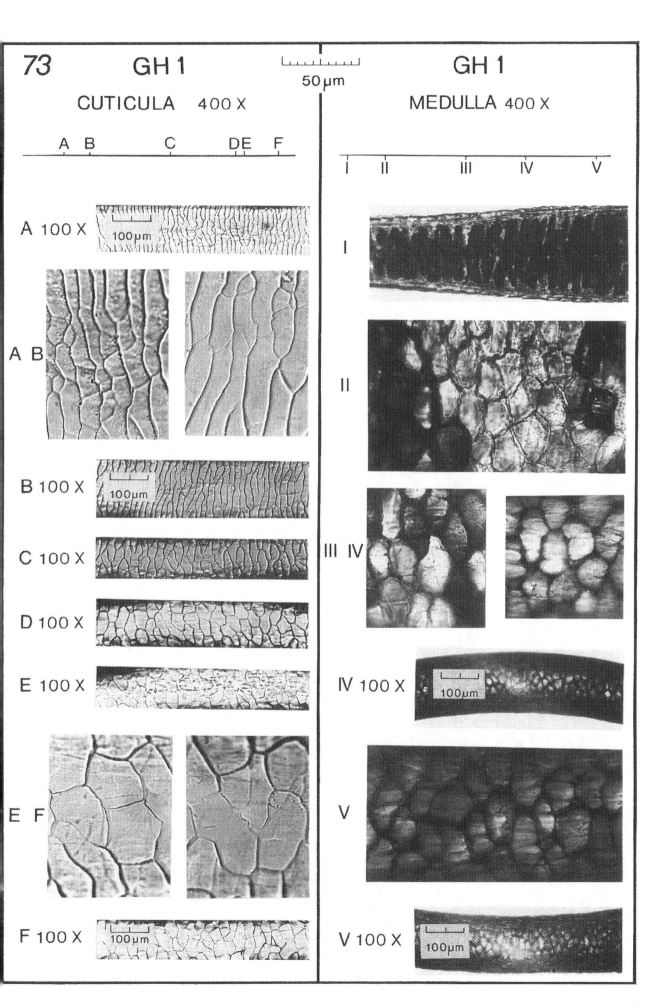

73   GH 1   50 µm   GH 1

CUTICULA 400 X   MEDULLA 400 X

A B   C   DE F   I   II   III   IV   V

A 100 X

A B

B 100 X

C 100 X

D 100 X

E 100 X

E F

F 100 X

I

II

III IV

IV 100 X

V

V 100 X

8 List of species

The species-have been classified into orders. The number before the species name has been used throughout the study. In this way the species concerned can be easily found in the atlas section. The species are listed successively with their scientific names and vernacular names in English, German, French, Dutch, and Danish.

Scientific name

INSECTIVORA
01 *Erinaceus europaeus* L.
02 *Sorex araneus* L.
03 *Sorex minutus* L.
04 *Neomys fodiens* (PENNANT)
05 *Neomys anomalus* CABRERA
06 *Crocidura russula* (HERMANN)
07 *Crocidura suaveolens* (PALLAS)
08 *Crocidura leucodon* (HERMANN)
09 *Talpa europaea* L.

CHIROPTERA
10 *Rhinolophus ferrumequinum* (SCHREBER)
11 *Rhinolophus hipposideros* (BECHSTEIN)
12 *Myotis mystacinus* KUHL
13 *Myotis brandtii* EVERSMANN
14 *Myotis emarginatus* E.GEOFFROY SAINT HILAIRE
15 *Myotis nattereri* KUHL
16 *Myotis bechsteinii* KUHL
17 *Myotis myotis* (BORKHAUSEN)
18 *Myotis daubentonii* KUHL
19 *Myotis dasycneme* BOIE
20 *Pipistrellus pipistrellus* (SCHREBER)
21 *Pipistrellus nathusii* (KEYSERLING & BLASIUS)
22 *Nyctalus noctula* (SCHREBER)
23 *Nyctalus leisleri* (KUHL)
24 *Eptesicus serotinus* (SCHREBER)
25 *Vespertilio murinus* L.
26 *Barbastella barbastellus* (SCHREBER)
27 *Plecotus auritus* (L.)
28 *Plecotus austriacus* (FISCHER)

RODENTIA
29 *Cricetus cricetus* (L.)
30 *Clethrionomys glareolus* (SCHREBER)
31 *Arvicola terrestris* (L.)
32 *Ondatra zibethicus* (L.)
33 *Pitymys subterraneus* (DE SELYS-LONGCHAMPS)

34 *Microtus arvalis* (PALLAS)
35 *Microtus agrestis* (L.)
36 *Microtus oeconomus* (PALLAS)
37 *Micromys minutus* (PALLAS)
38 *Apodemus sylvaticus* L.
39 *Apodemus flavicollis* (MELCHIOR)
40 *Apodemus agrarius* (PALLAS)
41 *Rattus norvegicus* (BERKENHOUT)
42 *Rattus rattus* (L.)
43 *Mus musculus* (L.)
44 *Glis glis* (L.)
45 *Muscardinus avellanarius* (L.)
46 *Eliomys quercinus* (L.)
47 *Castor fiber* L.
48 *Myocastor coypus* (MOLLINA)
49 *Sciurus vulgaris* L.
50 *Sciurus carolinensis* GMELIN
51 *Tamias sibiricus* (LAXMANN)

LAGOMORPHA
52 *Lepus europaeus* PALLAS
53 *Lepus timidus* L.
54 *Oryctolagus cuniculus* (L.)

CARNIVORA
55 *Vulpes vulpes* (L.)
56 *Canis familiaris* L.
57 *Nyctereutes procyonoides* (GRAY)
58 *Procyon lotor* (L.)
59 *Mustela erminea* L.
60 *Mustela nivalis* L.
61 *Mustela putorius* L.
62 *Mustela vison* (SCHREBER)
63 *Martes martes* (L.)
64 *Martes foina* (ERXLEBEN)
65 *Meles meles* (L.)
66 *Lutra lutra* (L.)
67 *Felis silvestris* (SCHREBER)
68 *Felis catus* L.

List of species

ARTIODACTYLA
69 *Sus scrofa* L.
70 *Cervus dama* L.
71 *Cervus elaphus* L.
72 *Capreolus capreolus* (L.)
73 *Ovis musimon* (PALLAS)

English

INSECTIVORA
01 European hedgehog
02 Common shrew
03 Pygmy shrew
04 Water shrew
05 Miller's water shrew
06 White-toothed shrew
07 Lesser white-toothed shrew
08 Bicoloured shrew
09 Mole

CHIROPTERA
10 Greater horseshoe bat
11 Lesser horseshoe bat
12 Whiskered bat
13 Brandt's bat
14 Geoffroy's bat
15 Natterer's bat
16 Bechstein's bat
17 Large mouse-eared bat
18 Daubenton's bat
19 Pond bat
20 Pipistrelle
21 Nathusius's pipistrelle
22 Noctule
23 Leisler's bat
24 Serotine
25 Parti-coloured bat
26 Barbastelle
27 Long-eared bat
28 Grey long-eared bat

RODENTIA
29 Common hamster
30 Bank vole
31 Ground vole
32 Muskrat
33 Pine vole
34 Common vole
35 Short-tailed vole
36 Root vole
37 Harvest mouse
38 Wood mouse
39 Yellow-necked mouse
40 Striped field mouse
41 Brown rat
42 Black rat
43 House mouse
44 Fat dormouse
45 Dormouse
46 Garden dormouse
47 Beaver
48 Coypu
49 Red squirrel
59 Grey squirrel
51 Asiatic chipmunk

LAGOMORPHA
52 Brown hare
53 Blue hare
54 Rabbit

CARNIVORA
55 Red fox
56 Dog
57 Raccoon dog
58 Raccoon
59 Stoat
60 Weasel
61 Polecat
62 American mink
63 Pine marten
64 Beech marten
65 Badger
66 Otter
67 Wild cat
68 Domestic cat

ARTIODACTYLA
69 Wild boar
70 Fallow deer
71 Red deer
72 Roe deer
73 Mouflon

German

INSECTIVORA
01 Igel
02 Waldspitzmaus
03 Zwergspitzmaus
04 Wasserspitzmaus
05 Sumpfspitzmaus
06 Hausspitzmaus
07 Gartenspitzmaus
08 Feldspitzmaus
09 Maulwurf

CHIROPTERA
10 Grosshufeisennase
11 Kleinhufeisennase
12 Kleine Bartfledermaus
13 Grosse Bartfledermaus
14 Wimperfledermaus
15 Fransenfledermaus
16 Bechsteinfledermaus
17 Mausohr
18 Wasserfledermaus
19 Teichfledermaus
20 Zwergfledermaus
21 Rauhautfledermaus
22 Abendsegler
23 Kleinabendsegler
24 Breitflügelfledermaus
25 Zweifarbfledermaus
26 Mopsfledermaus
27 Braunes Langohr
28 Graues Langohr

RODENTIA
29 Hamster
30 Rötelmaus
31 Schermaus

32 Bisamratte
33 Kleinwühlmaus
34 Feldmaus
35 Erdmaus
36 Sumpfmaus
37 Zwergmaus
38 Waldmaus
39 Gelbhalsmaus
40 Brandmaus
41 Wanderratte
42 Hausratte
43 Hausmaus
44 Siebenschläfer
45 Haselmaus
46 Gartenschläfer
47 Biber
48 Nutria
49 Eichhörnchen
50 Grauhörnchen
51 Burunduk

LAGOMORPHA
52 Feldhase
53 Schneehase
54 Wildkaninchen

CARNIVORA
55 Rotfuchs
56 Hund
57 Marderhund
58 Waschbär
59 Hermelin
60 Mauswiesel
61 Iltis
62 Mink
63 Baummarder
64 Steinmarder
65 Dachs
66 Fischotter
67 Wildkatze
68 Hauskatze

ARTIODACTYLA
69 Wildschwein
70 Damhirsch
71 Rothirsch
72 Reh
73 Mufflon

French

INSECTIVORA
01 Hérisson d'Europe
02 Musaraigne carrelet
03 Musaraigne pygmée
04 Musaraigne aquatique
05 Musaraigne de Miller
06 Musaraigne musette
07 Musaraigne des jardins
08 Musaraigne bicolore
09 Taupe d'Europe

CHIROPTERA
10 Grand Rhinolphe fer à cheval
11 Petit Rhinolphe fer à cheval
12 Vespertilion à moustaches

13 Vespertilion de Brandt
14 Vespertilion à oreilles échancrées
15 Vespertilion de Natterer
16 Vespertilion de Bechstein
17 Vespertilion murin
18 Vespertilion de Daubenton
19 Vespertilion des marais
20 Pipistrelle
21 Pipistrelle de Nathusius
22 Noctule
23 Noctule de Leisler
24 Grande serotine
25 Serotine bicolore
26 Barbastelle d'Europe
27 Oreillard brun
28 Oreillard gris

RODENTIA
29 Grand hamster
30 Campagnol roussâtre
31 Campagnol terrestre
32 Rat musqué
33 Campagnol souterrain
34 Campagnol des champs
35 Campagnol agreste
36 Campagnol nordique
37 Rat des moissons
38 Mulot sylvestre
39 Mulot à collier roux
40 Mulot rayé
41 Surmulot
42 Rat noir
43 Souris grise
44 Loir
45 Muscardin
46 Lérot
47 Castor
48 Ragondin
49 Ecureuil d'Europe
50 Ecureuil gris
51 Ecureuil terrestre rayé

LAGOMORPHA
52 Lièvre brun
53 Lièvre variable
54 Lapin de garenne

CARNIVORA
55 Renard roux
56 Chien domestique
57 Chien viverrin
58 Raton laveur
59 Hermine
60 Belette
61 Putois
62 Vison d'Amérique
63 Martre des pins
64 Fouine
65 Blaireau d'Europe
66 Loutre
67 Chat sauvage
68 Chat domestique

ARTIODACTYLA
69 Sanglier
70 Daim

71 Cerf rouge
72 Chevreuil
73 Mouflon

Dutch

INSECTIVORA
01 Egel
02 Bosspitsmuis
03 Dwergspitsmuis
04 Waterspitsmuis
05 Millsre waterspitsmuis
06 Huisspitsmuis
07 Tuinspitsmuis
08 Veldspitsmuis
09 Mol

CHIROPTERA
10 Grote hoefijzerneus
11 Kleine hoefijzerheus
12 Baardvleermuis
13 Brandts vleermuis
14 Ingekorven vleermuis
15 Franjestaart
16 Bechsteins vleermuis
17 Vale vleermuis
18 Watervleermuis
19 Meervleermuis
20 Dwergvleermuis
21 Nathusius' dwergvleermuis
22 Rosse vleermuis
23 Bosvleermuis
24 Laatvlieger
25 Tweekleurige vleermuis
26 Mopsvleermuis
27 Grootoorvleermuis
28 Grijze grootoorvleermuis

RODENTIA
29 Hamster
30 Rosse woelmuis
31 Woelrat
32 Muskusrat
33 Ondergrondse woelmuis
34 Veldmuis
35 Aardmuis
36 Noordse woelmuis
37 Dwergmuis
38 Bosmuis
39 Grote bosmuis
40 Brandmuis
41 Bruine rat
42 Zwarte rat
43 Huismuis
44 Zevenslaper
45 Hazelmuis
46 Eikelmuis
47 Bever
48 Beverrat
49 Eekhoorn
50 Grijze eekhoorn
51 Siberische grondeekhoorn

LAGOMORPHA
52 Haas
53 Sneeuwhaas

54 Konijn

CARNIVORA
55 Vos
56 Hond
57 Wasbeerhond
58 Wasbeer
59 Hermelijn
60 Wezel
61 Bunzing
62 Amerikaanse nerts
63 Boommarter
64 Steenmarter
65 Das
66 Otter
67 Wilde kat
68 Huiskat

ARTIODACTYLA
69 Wild zwijn
70 Damhert
71 Edelhert
72 Ree
73 Moeflon

Danish

INSECTIVORA
01 Pindsvin
02 Alm. spidsmus
03 Dværgspidsmus
04 Vandspidsmus
05 Millers vandspidsmus
06 Husspidsmus
07 Havespidsmus
08 Markspidsmus
09 Muldvarp

CHIROPTERA
10 Stor hesteskouæse
11 Lille hesteskouæse
12 Skælflagermus
13 Brandts flagermus
14 Geoffroys flagermus
15 Frynse flagermus
16 Bechsteins flagermus
17 Stor museøre
18 Vand Hagermus
19 Damflagermus
20 Dværgflagermus
21 Trold Hagermus
22 Brun flagermus
23 Leislers flagermus
24 Syd flagermus
25 Skimmelflagermus
26 Bredøret flagermus
27 Langøret flagermus
28 Grå langøret flagermus

RODENTIA
29 Hamster
30 Rødmus
31 Mosegris/Vandrotte
32 Bisamrotte
33 Kortøret markmus
34 Sydmarkmus

221

Part III: Atlas

35 Alm. markmus
36 Markrotte
37 Dværgmus
38 Skormus
39 Halsbåndmus
40 Brandmus
41 Vandrerotte
42 Sortrotte
43 Husmus
44 Sysrorer
45 Hasselmus
46 Havesyvsover
47 Bæver
48 Sumpbæver
49 Egern
50 Grå egern
51 Sibirsk jordegern

LAGOMORPHA
52 Hare
53 Snehare
54 Vildkanin

CARNIVORA
55 Ræv
56 Huud
57 Mårhund
58 Vaskebjørn
59 Lækat/Hermelin
60 Brud
61 Ilder
62 Mink
63 Skormår
64 Husmår
65 Grævling
66 Odder
67 Vildkat
68 Kat

ARTIODACTYLA
69 Vildsvin
70 Dådyr
71 Kronhjort
72 Rådyr
73 Muflon

9 References

Becker, K. (1952). Haarwechselstudien an Wanderratten (*Rattus norvegicus* Erxl.). *Biol. Zentralbl.* **71**: 626–40, Berlin (cited by Stein, 1960).

Benedict, F.A. (1957). Hair structure as a generic character in bats. *Univ. Calif. Publ. Zool.* **59**: 285–548.

Bereiter-Hahn, J., Matoltsy, A.G. & Richards, K.S. (1984).* *Biology of the Integument*, Vol. 2, Vertebrates. Springer-Verlag Berlin, Heidelberg, New York, Tokyo.

Brewster, (1837). *Treatise on the microscope* (cited by Tupinier, 1973).

Brunner, H. & Coman, B.J. (1974). *The Identification of Mammalian Hair*. Inkata Press, Melbourne.

Day, M.G. (1965). Identification of hair and feather remains in the gut and faeces of stoats and weasels. *J. Zool.* **148**: 201–17.

Debrot, S., Fivaz, G., Mermod C. & Weber J.M. (1982). *Atlas des Poils de Mammifères d'Europe*. Institut de Zoologie de l'Université de Neuchâtel.

Dziurdzik, B. (1973). Key to the identification of hairs of mammals from Poland (Polish). *Acta Zool. Cracov.* **18**: 73–91.

Dziurdzik B. (1978). Histological structure of hair in the *Gliridae* (*Rodentia*). *Acta Zool. Cracov.* **23**: 1–10.

Ford, J.E. & Simmens, S.C. (1959). Fibre section cutting by the plate method. *J. Text. Inst. Proc.* **50**: 148–58.

Haitlinger, R. (1968a). Seasonal variation of pelage in representatives of the genus *Apodemus* Kaup, 1829, found in Poland. *Zool. Pol.* **18**: 329–45.

Haitlinger, R. (1968b). Comparative studies on the morphology of hair in representatives of the genus *Apodemus* Kaup, 1829, found in Poland. *Zool. Pol.* **18**: 347–80.

Hausman, L.A. (1920). Structural characteristics of the hair of mammals. *Am. Nat.* **54**: 496–523.

Hausman, L.A. (1924). Further studies of the relationships of the structural characters of mammalian hair. *Am. Nat.* **58**: 544–57.

Hausman, L.A. (1930). Recent studies of hair structure relationships. *Scient. Monthly.* **30**: 258–77.

Hausman, L.A. (1932). The cortical fusi of mammalian hair shafts. *Am. Nat.* **66**: 461–70.

Hausman, L.A. (1944). Applied microscopy of hair. *Scient. Monthly.* **59**: 195–202.

Hutterer, R. & Hürter T. (1981). Adaptive Haarstrukturen bei Wasserspitzmäusen (*Insectivora, Soricinae*). *Z. Säugetierk.* **46**: 1–11.

Keller, A. (1978). Determination des mammifères de la Suisse par leur pelage I. *Talpidae* et *Soricidae*. *Rev. Suisse Zool.* **85**: 758–61.

Keller, A. (1980). Determination des mammifères de la Suisse par leur pelage II.

*Not mentioned in the text.

References

Diagnose des familles. III. *Lagomorpha, Rodentia* (partim). *Rev. Suisse Zool.* **87**: 781–96.

Keller, A. (1981a). Determination des mammifères de la Suisse par leur pelage IV. *Cricetidae* et *Muridae. Rev. Suisse Zool.* **88**: 463–73.

Keller, A. (1981b). Determination des mammifères de la Suisse par leur pelage V. *Carnivora. Artiodactyla. Rev. Suisse Zool.* **88**: 803–20.

Lochte, T. (1938). Atlas der menschlichen und tierischen Haare. *Verlag Dr. Schops,* Leipzich.

Lyne, A.G. (1966). The development of hair follicles. *Austr. J. Sci.* **28**: 370–7.

Mathiak, H.A. (1938a). A key to hairs of the mammals of southern Michigan. *J. Wildl. Mgmt.* **2**: 251–68.

Mathiak, H.A. (1938b). A rapid method of cross-sectioning mammalian hairs. *J. Wildl. Mgmt.* **2**: 162–4.

Quekett, J. (1844). On the structure of bat's hair. *Trans. Microsc. Soc. Lond.* **1**: 58–62. (cited by Tupinier, 1973).

Short, H.L. (1978). Analysis of cuticular scales on hairs using the scanning electron microscope. *J. Mammal.* **59**: 261–7.

Stein, G.H.W. (1960). Zum Haarwechsel der Feldmaus, *Microtus arvalis* (Pallas, 1779) und weitere *Muroidea. Acta Ther.* **3**: 27–44.

Tupinier, Y. (1973). Morphologie des poils de Chiroptères d'Europe occidentale par étude du microscope électronique à balayage. *Rev. Suisse Zool.* **80**: 635–53.

Tupinier, Y. (1974). Morphologie des poils de Chiroptères d'Europe *Myotis brandtii* (Eversmann, 1845). *Rev. Suisse Zool.* **81**: 41–3.

Twigg, G.I. (1975). Finding mammals – their signs and remains. In: Techniques in mammalogy. *Mamm. Rev.* **5**: 77–8.

Vogel, P. & Köpchen B. (1978). Besondere Haarstrukturen der *Soricidae* (*Mammalia, Insectivora*) und ihre taxonomische Deutung. *Zoomorphology.* **89**: 47–56.

Wildman, A.B. (1954). The microscopy of animal textile fibres. *Leeds: Wool Ind. Res. Assoc.*

Wildman, A.B. (1961). The identification of animal fibres. *J. Forens. Sci. Soc.* **1**: 1–8.

Williams C.S. (1938). Aids to the identification of mole and shrew hairs with general comments on hair structure and hair determination. *J. Wildl. Mgmt.* **2**: 239–50.

Williamson, V.H.H. (1951). Determination of hairs by impressions. *J. Mammal.* **32**: 80–5.